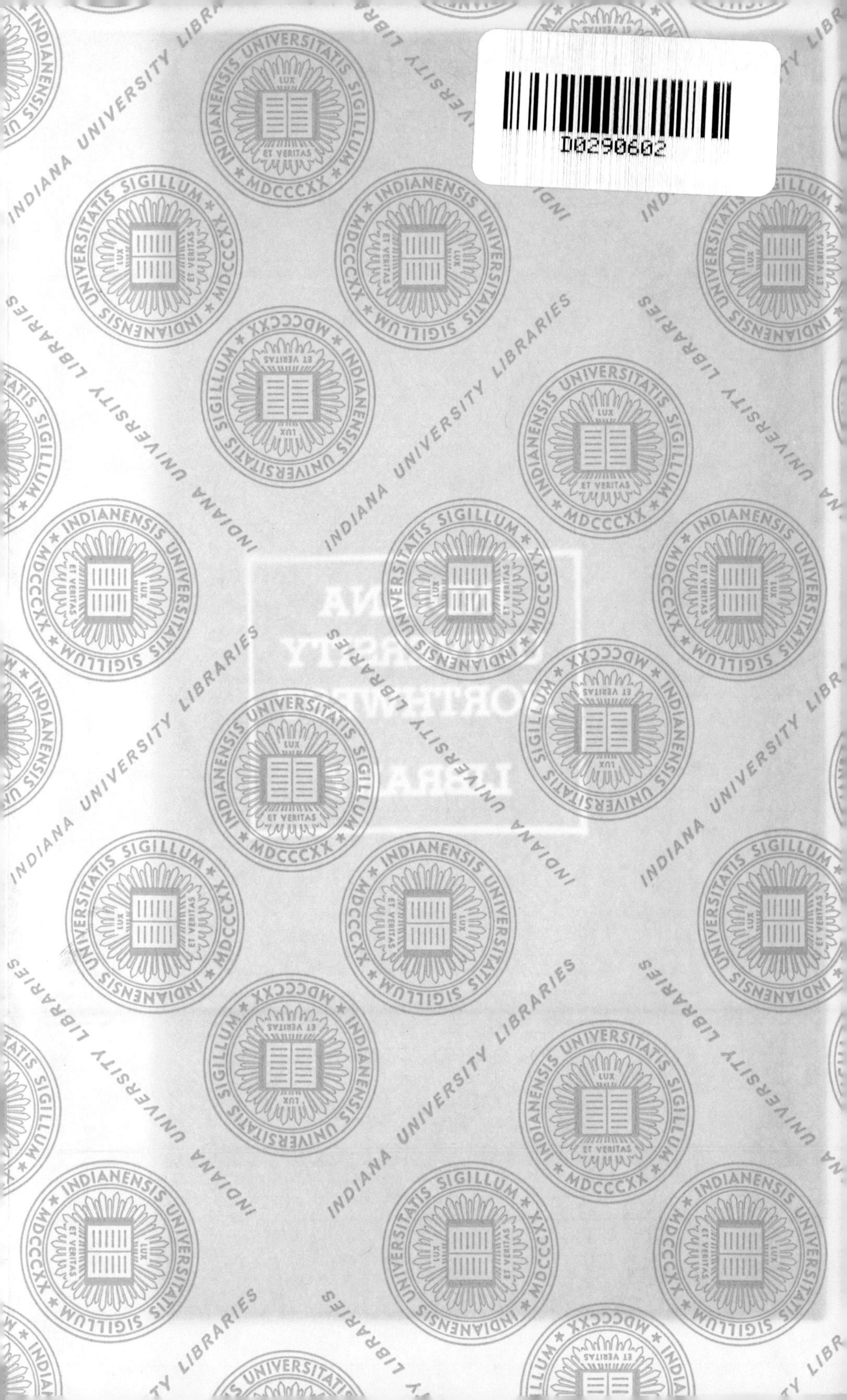
D0290602

INDIANA
UNIVERSITY
NORTHWEST
LIBRARY

QC 21.2
.D86

Physics for Biologists

GEORGE DUNCAN

School of Biological Sciences
University of East Anglia
Norwich

A HALSTED PRESS BOOK

JOHN WILEY & SONS INC
NEW YORK

INDIANA
UNIVERSITY
LIBRARY

NORTHWEST

© 1975 Blackwell Scientific Publications

All rights reserved. No part of this publication may be reproduced, stored in a retrieval system, or transmitted, in any form or by any means, electronic, mechanical, photocopying, recording or otherwise without the prior permission of the copyright owner.

First published 1975

Library of Congress Cataloging in Publication Data

Duncan, George
Physics for biologists.

"A Halsted Press book."
1. Physics. I. Title.
QC21.2.D86 530'.02'4574 74-18621
ISBN 0-470-22568-8

Printed in Great Britain and
Published in the U.S.A. by
Halsted Press, a Division of
John Wiley & Sons, Inc.
New York.

ISBN: 0 470

Contents

Preface

It is my firm belief that a successful 'Physics for Biologists' course can only be given by biologists or at least by those who are sympathetic towards biology. Undergraduates today are mercifully unwilling to put up with the Syrup of Figs attitude to this part of their education; they no longer accept their physics medicine with the vague promise that although it may not taste good now, it is surely going to have some beneficial effect in the not too distant future. They want to see the immediate relevance of the subject *now* and it is for these reasons that I have written this text.

I also hope that it will give a further insight into physics to those biology teachers who are giving a course and an insight into biology to those teachers in traditional Physics (or Natural Philosophy) departments who are given the onerous task of entertaining restive biologists. Most of all, however, it is to you restive biologists that this text is addressed. Ideally, it would take the form of a short explanation of the few basic principles of physics involved, followed by a long, a very long, reading list of original and review articles illustrating the biological applications of these principles. However, I realise that time is in short supply, that there are physical chemistry, organic chemistry, mathematics, statistics, computing and even biology lectures to attend as well, so I have here interwoven the principles and applications. At the end of each chapter I have listed some research articles and it is important that you should read at least one or two of those in a field that specially interests you, so that you can see how much more alive the subject of *Biophysics* becomes on the battlefield itself.

As biologists we are continually faced with the situation where we have to describe a biological system in basic physical terms in order to learn more about the underlying physiological mechanisms. For example, Bennet-Clark and Lucey working in Edinburgh used hundreds of metres of high-speed photographic film in order to analyse the jump of the human flea. In this way they were able to time the jump and to measure the maximum height reached. When they then worked out the overall energy consumption for the jump they came to the startling conclusion that the flea's muscles simply could not supply this energy in the time available and this led them to propose an entirely new jumping mechanism. In the field of Phloem Translocation there is a similar problem because a relatively simple computation shows that conventional driving forces (e.g. osmotic pressure) are insufficient to transport water and solutes down the phloem at the rates that are normally found. This has stimulated a search for more unconventional mechanisms, which in turn has led to a reinvestigation of plant ultrastructure.

Both of these examples (along with many others) are set as problems and the answers, given in full, should be read as an integral part of the text. Problems of a revision nature are also set in the appropriate section and fully-worked answers are given at the back. The problems are presented in this way rather than as an indigestible lump at the end of each chapter so that problem solving can be seen as a valuable and integral part of a biophysics course rather than a chore to be sweated over in the tedium of the separate tutorial.

October 1974

Acknowledgments

I wish to thank the students and faculty of the School of Biological Sciences of the University of East Anglia for many discussions on all aspects of this text and in particular I would like to thank Drs D. J. Aidley, E. J. A. Lea, P. Smith and Mr R. B. Clark for many stimulating discussions on specific issues. I am indebted to Dr M. Jarman of the University of Bristol and Dr P. C. Croghan and Mr P. B. Pynsent of U.E.A. for their constructive criticisms of the whole text. I also gratefully acknowledge the help and encouragement of Mr R. Campbell of Blackwell Scientific Publications. Finally, I would also like to thank Mrs Sue Mayer for typing the many manuscript drafts.

Chapter 1
Basic Physics: Mechanical Properties of Matter

1.1 Introduction

All of us who have undergone the transition to the metric system in Britain will understand the need for a universally acceptable system for the units of measurement. In the rush and bustle of a supermarket it is more than annoying to have to assess the relative merits of one can of peaches costing 10p for $15\frac{1}{2}$ oz with another costing 10p for 0.5 kg. Similarly, in biology or indeed in any other science, participation in any discussion is hampered when those involved are familiar with different systems of units. In this book, the language will be that of the Système International (SI), but from time to time a mention will be made of other systems so that the older literature of biology can be read with some ease.

1.2 Defined Quantities: Mass, Length, and Time

The foundation of any physical science rests on the definition of the fundamental quantities, mass, length, and time (m, l, and t). The universally accepted standard of *mass* is at present the mass of a lump of platinum in Paris but the secondary definition based on the mass of the ^{12}C isotope of carbon (see Chapter 10, p. 171 for a definition of *isotope*) will probably soon become the accepted standard. In the SI system the unit of mass is the kilogram and on the atomic mass scale, one kilogram is defined as the mass of 5.02×10^{25} atoms of ^{12}C. *Length* is now defined in terms of the wavelength (Chapter 7) of the orange line of krypton and the SI unit is the metre. *Time* is defined in terms of the vibration of the caesium atom and the SI unit is the second. The correct abbreviations of the many SI units are given in Appendix I.

1.3 Derived Quantities: Scalars and Vectors

There are many physical quantities derived from the above fundamentals. The *velocity* (v) of a body is defined as the rate of change of position of the body with respect to time and it has dimensions lt^{-1} and units $m s^{-1}$. Mass, length, and time are examples of *scalar* quantities, which means they have magnitude only, whereas velocity is an example of a *vector* quantity having both magnitude and direction.

A vector (AB in Fig. 1.1) is usually represented by an arrow, the length of which gives a measure of the magnitude of the vector and the direction of the arrow indicates the direction of the vector. In this text vectors will be denoted by letters in bold type.

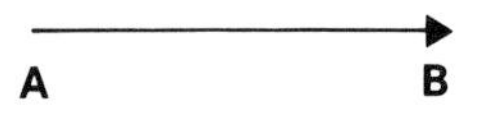

Fig. 1.1. The vector **AB** has both magnitude and direction.

Vectors are subject to the laws of vector algebra (Fig. 1.2). A simple law worth remembering is that vectors are added from tail to head (Fig. 1.3). The vector **AC** is called the *resultant* of **AB** and **BC** and represents the net effect of the other two vectors.

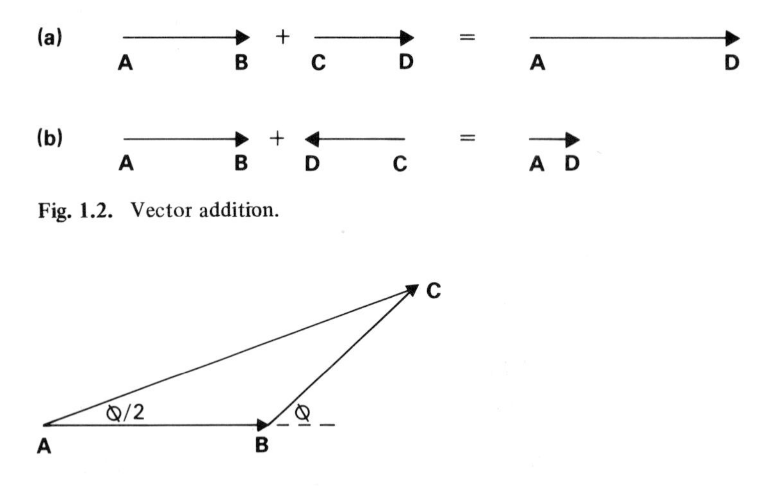

Fig. 1.2. Vector addition.

Fig. 1.3. The vectors **AB** and **BC** are ϕ radians out of alignment. It is left as an exercise to show that when **AB** equals **BC**, then the resultant is $\phi/2$ radians out of alignment with both **AB** and **BC**. The algebraic notation is **AB** + **BC** = **AC**.

Any one vector can be represented by two vectors at right angles to one another (Fig. 1.4). If the magnitude of **AB** is a, then the magnitudes of **AM** and **AL** are $a \cos \theta$ and $a \sin \theta$, respectively. **AM** and **AL** are said to be the components of **AB** and it is left as a simple exercise to show that when **AL** and **AM** are added tail to head, the resultant is **AB**. It is worth noting that θ is not unique; a vector can be resolved into components along any pair of axes.

In calculus notation (Ferrar, 1967, or any elementary calculus text), the rate of change of position s with respect to time t is written as

$$\mathbf{v} = ds/dt \qquad 1.1$$

The *momentum* of a body is defined as the product of its mass and velocity and is a vector quantity with dimensions mlt^{-1}.

Acceleration, also a vector, is defined as the rate of change of velocity with time

$$\mathbf{a} = d\mathbf{v}/dt = d^2\mathbf{s}/dt^2 \quad 1.2$$

Acceleration has dimensions lt^{-2} and units $\mathrm{m\,s^{-2}}$.

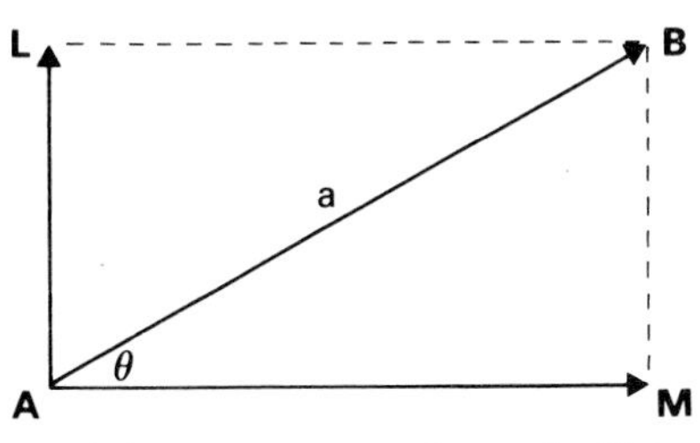

Fig. 1.4. Any vector can be resolved into two components at right angles. If the length of vector **AB** (the magnitude of a vector is often written |**AB**| or simply AB) is a, then magnitude of **AL** and **AM**, the components of the original vector, are $a \sin \theta$ and $a \cos \theta$ respectively.

1.4 Mass and Force: Newton's Laws

Sir Isaac Newton was the great English scientist who discovered the basic laws governing the interaction of forces and masses. His three Laws of Motion form the basis for the science of mechanics and they are summarized below.

(i) Every body continues in a state of rest or of uniform motion in a straight line unless it is compelled to change that state by an external force.

(ii) The rate of change of momentum of a body is equal to the resultant of all external forces exerted on the body.

(iii) When one body exerts a force on another, the second always exerts on the first a force, called the reaction force, which is equal in absolute magnitude, opposite in direction and has the same line of action.

Newton's second law can be summarized in the equation

$$\mathbf{F} = \frac{d}{dt}(m\mathbf{v}) \quad 1.3a$$

and as in most cases of biological interest the mass of the body remains constant while it is being acted on by the force (rockets in flight are an exception) then equation 1.3a simplifies to equation 1.3b

$$\mathbf{F} = m\mathbf{a} \quad 1.3b$$

In the SI system the basic unit of force is the *newton* which is the force required to give a mass of 1 kg an acceleration of $1\ \mathrm{m\,s^{-2}}$.

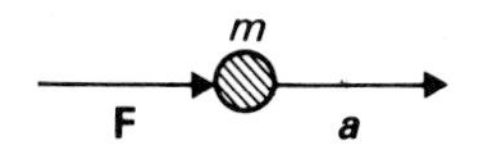

Fig. 1.5. Summary of Newton's second law. A vector force **F** acting on a body of mass m produces an acceleration **a**.

Because different observers can describe the motion of the same body quite differently, a *frame of reference* must be introduced when describing the motion. For example, consider the hypothetical case of a passenger standing on frictionless roller skates in the corridor of a train that is accelerating away from a station (Fig. 1.6). His fellow passengers will note that he is accelerating backwards relative to them and so would say that he is experiencing a force. An observer standing at rest on the station would note that the roller-skated passenger remained more or less at rest and so would conclude that no force was acting on him.

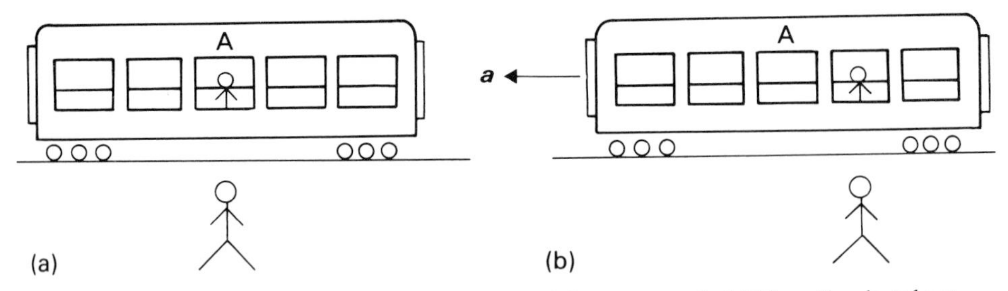

Fig. 1.6. Frames of reference (two views of the same event). (a) The roller-skated passenger is standing opposite window A when the train is at rest. (b) The train moves off with an acceleration a. Other passengers on the train will observe him moving backwards relative to them with an acceleration a, whereas an observer on the platform would say that he has scarcely moved during this time.

Newton's first law cannot hold in both cases and it is in fact in the frame of reference fixed within the train that the law does not hold. There is no real backwards force on the roller-skated passenger, simply an apparent force. However, in order for Newton's first law to hold for an observer accelerating with the train, a force must be added to the passenger in a direction opposite to the acceleration of the moving frame. This force is sometimes called the *inertial* or *fictional* force.

1.5 Gravitational and Inertial Forces

Newton also discovered the Law of Universal Gravitation which forms the basis of theoretical astronomy. It states that every body in the universe attracts every other body with a force **F** that is directly proportional to the product of their masses and inversely proportional to the square of the distance between them (Fig. 1.7). The magnitude of the force is given by the equation

$$F = G\frac{m_1 m_2}{r^2} \qquad 1.4$$

where G is the *gravitational constant* and is equal to 6.7×10^{-11} Nm^2 kg^{-2}.

Gravitational forces are relatively weak compared to electrical (Chapter 9, p. 138) and nuclear (Chapter 10, p. 170) forces and are only significant when very large masses, e.g. planets, are involved.

The *weight* of a body is defined as the resultant gravitational force exerted on the body by all other bodies in the universe. The gravitational force of attraction on a body at the surface of the earth is such as to cause it to accelerate at about 9.8 m s^{-2} towards the centre of the earth. (The gravitational acceleration is usually denoted by the letter g.) The resultant force due to gravity on a mass of 1 kg is therefore 9.8 N or

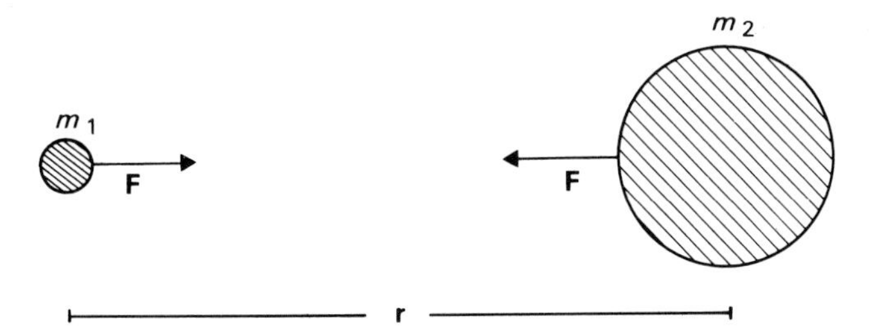

Fig. 1.7. Newton's Law of Universal Gravitation states that the force which m_1 exerts on m_2 is equal and opposite to that which m_2 exerts on m_1. The magnitude of each force is given by equation 1.4.

1 kg weight. The weight of a body in fact varies from point to point on the earth's surface as the gravitational acceleration varies due to inhomogeneities in the earth's composition and because the earth is not perfectly spherical. The mass of a body on the other hand does not vary as it is determined by comparison with a standard mass on a balance.

The gravity-like nature of inertial forces is obvious in accelerating lifts (Fig. 1.8). (*a*) When the lift is accelerating upwards a person standing on a spring-balance weighing machine will note that his weight will have increased by an amount *ma* as the inertial force is in the same direction as the gravitational force. (*b*) Should the lift cable break the person will note that according to the machine he will have become

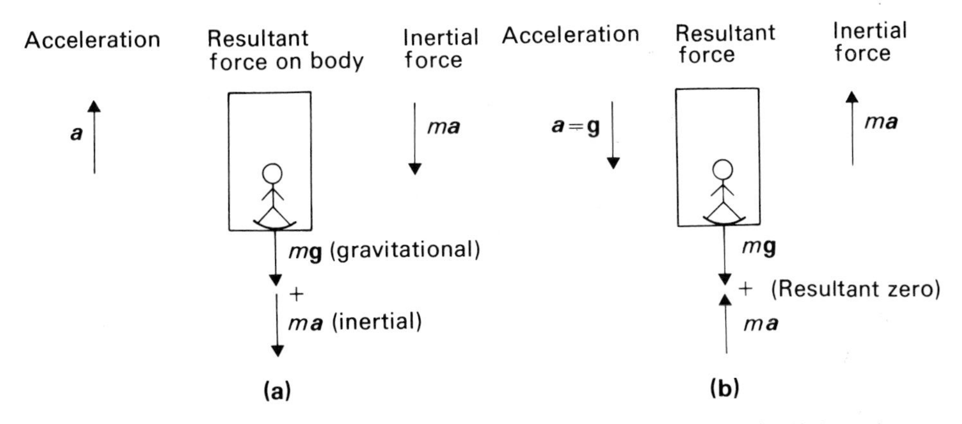

Fig. 1.8. (a) The inertial force adds to the gravitational force when the lift is accelerating upwards. (b) The person will appear weightless when the lift cable breaks.

weightless. If the person had, however, been standing on an oversized chemical balance (one which compares his mass against a standard) his apparent mass would remain unchanged in the accelerated frames. In the case where the lift was moving with constant velocity, the weight as measured by the spring balance would remain unchanged because the frame of reference moving with the lift would be an inertial one.

1.6 Equations of Motion

From equation 1.2 several equations of motion can be derived that can be applied to situations when a mass moves under a constant acceleration, e.g. a mass moving in a gravitational field. Suppose we wish to know the velocity of the mass at any time t, then integrating equation 1.2

$$\int_{v_o}^{v_t} dv = \int_{t_o}^{t} a\,dt$$

hence

$$v_t - v_o = at - at_o \qquad 1.5$$

where v_o and v_t are the velocities at times t_o and t. When $t_o = 0$, then

$$v_t = v_o + at \qquad 1.6a$$

Equation 1.6a is actually a vector equation and is only true when the vectors representing v_o and a are in the same direction (Fig. 1.9a).

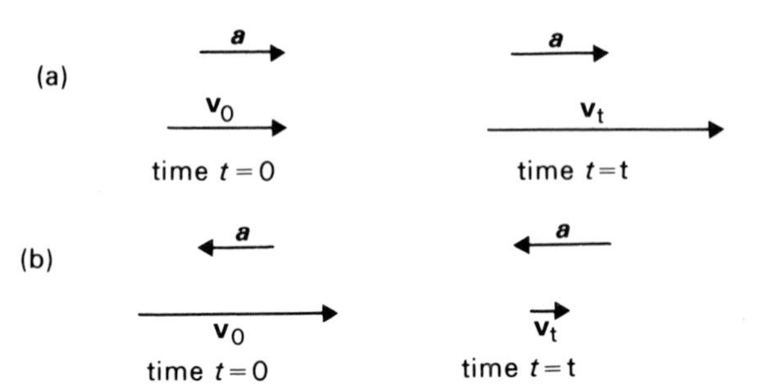

Fig. 1.9. (a) Equation $v_t = v_o$ + at describes the motion after a time t from zero. (b) Equation $v_t = v_o - at$ describes the motion.

When the vectors are in opposite directions (Fig. 1.9b) the equation takes the form

$$v_t = v_o - at \qquad 1.6b$$

Equation 1.6a can be expressed in the form

$$ds/dt = v_o + at$$

and integrating once more

$$s = v_o t + \tfrac{1}{2} at^2 \qquad 1.7a$$

when initial velocity and acceleration vectors are in the same direction (Fig. 1.10a), and

$$s = v_o t - \tfrac{1}{2} at^2 \tag{1.7b}$$

when they are in opposite directions (Fig. 1.10b). s is the distance travelled after a time t.

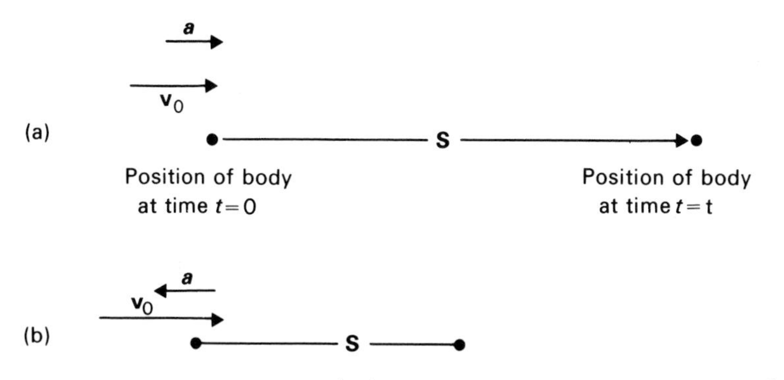

Fig. 1.10. (a) Equation $s = v_o t + \frac{1}{2}at^2$ describes motion. (b) Equation $s = v_o t - \frac{1}{2}at^2$ describes motion.

Combining equations 1.6a and 1.7a gives

$$v_t^2 = v_o^2 + 2as \tag{1.8a}$$

Combining 1.6b and 1.7b gives

$$v_t^2 = v_o^2 - 2as \tag{1.8b}$$

Problem 1.1

A salmon jumps 5 m to clear a waterfall. With what initial velocity did it leave the water? (Fig. 1.11)

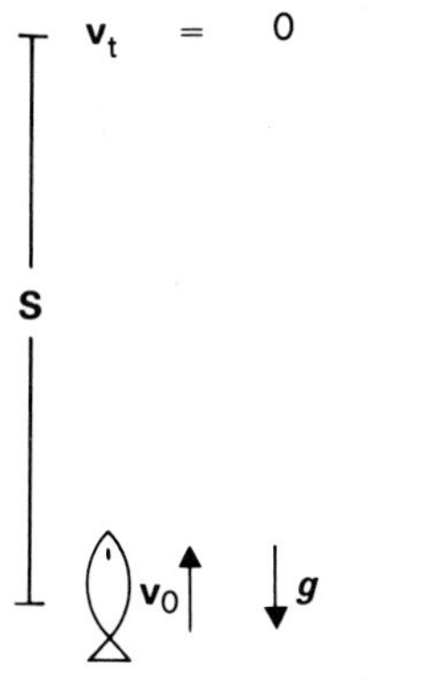

Fig. 1.11. (See text for explanation)

Answer
Apply equation 1.8b: $s = 5\,\text{m}$, $v_t = 0$, and $a = 9.8\,\text{m s}^{-2}$. This gives a result of approximately $10\,\text{m s}^{-1}$, or just over 20 m.p.h., for the vertical component of the take-off velocity.

1.7 Work and Energy

Forces, as we have already seen, are agents that are capable of accelerating masses but they can also be defined as the instruments by means of which the energy of a body is either increased or decreased. For example, a billiard ball lying at rest has its energy increased by hitting it with a billiard cue. The ball then gradually loses its energy again by frictional forces exerted on it by the table and also when it gives up some of its energy to other balls on the table.

When forces act on a body and the body moves, then work is done by these forces. The work done as a body moves through a distance s is the product of the component of the force along s and the vector displacement s (Fig. 1.12), i.e.

$$\text{work} = F\cos\theta s \tag{1.9}$$

The expression $F\cos\theta s$ is the scalar product of the two vectors F and s and is written $\mathbf{F}.\,\mathbf{s}$ (Appendix II). Note that the displacement is a vector quantity, as the direction in which the body moves is obviously important.

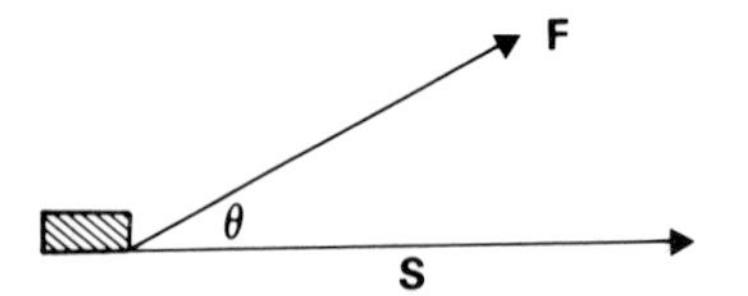

Fig. 1.12. A force **F** acts, the body actually moves along s, and work is done. This diagram could represent the work done in dragging a heavy object along a floor. Only the component **F** cos θ acts along s and so the work done is given by Fs cos θ. The vertical component **F** sin θ contributes nothing to the work done.

Work and energy have the same dimensions ml^2t^{-2} and both are scalar quantities. The SI unit of energy is the *joule* (J) which is defined as the work done when a force of one newton moves its point of application through a distance of 1 metre.
There are two forms of energy.

(i) *Potential Energy* is accumulated in a system as a result of previous work being done on the system. It is energy stored in chemical bonds for example, or energy stored in a body that has been moved some distance above the surface of the earth. The stored energy is released once the bond is broken or the body is returned to the surface.

(ii) *Kinetic Energy* is possessed by a moving body and is defined by the expression

$$KE = \tfrac{1}{2} mv^2 \qquad 1.10$$

Consider the salmon problem again (Fig. 1.13). Equation 1.8b gives

$$v_t^2 - v_o^2 = -2gs$$

and multiplying both sides by $\tfrac{1}{2}m$ and rearranging

$$\tfrac{1}{2}m\, v_o^2 - \tfrac{1}{2}m\, v_t^2 = mgs$$

Now the work done against gravity by the salmon in jumping a distance s is mgs which is the increase in potential energy and this is equal to the change in the salmon's kinetic energy. This illustrates the conservation of energy theorem, as the potential energy gained equals the kinetic energy lost. If there had been frictional forces present, e.g. air resistance, then this would not be true as part of the initial kinetic energy of the salmon would be used to overcome these forces. This energy would be lost to the body and so would not be converted into potential energy.

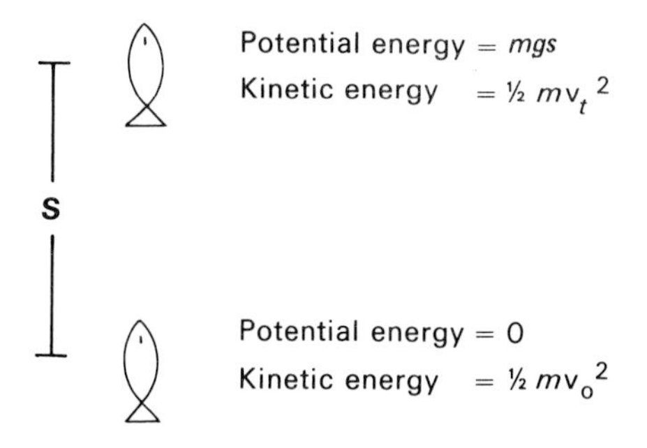

Fig. 1.13. (See text for explanation.)

The salmon moving solely under the influence of gravity represents a *conservative* system, whereas when dissipative forces are present the system is said to be *non-conservative*. A book pushed along a table-top is an example of a non-conservative system (Fig. 1.14).

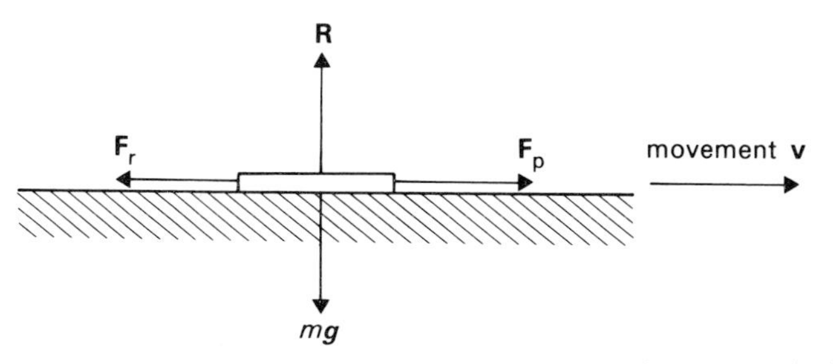

Fig. 1.14. A book pushed along a table-top moves with constant velocity under the action of four forces.

Heat is also a form of energy. The kinetic energy of the molecules making up a body determines the heat energy of the body and in the SI system the unit of heat energy is the joule (J). However, a unit sometimes used is the *calorie* (cal) and this is the heat required to raise 1 g of water by 1°C. Experimentally it has been found that

$$1 \text{ cal} = 4.2 \text{ J}$$

The kilocalorie is also much used, especially in metabolic rates which are usually quoted in kcal day^{-1}.

Power is the rate at which energy is expended and it has units of force x distance x time^{-1}. If the instantaneous velocity **v** of a body moving under the influence of a force **F** is known, then the instantaneous power developed P is given by

$$P = \mathbf{F}.\mathbf{v} \qquad 1.11a$$

The average power developed over a time is given by

$$P_{avge} = \frac{1}{t}\int_{o}^{t} \mathbf{F}.\mathbf{v}\, dt \qquad 1.11b$$

The basic unit of power is the *watt* and

$$1 \text{ W} = 1 \text{ J s}^{-1}$$

The inefficiency of man as a mechanical device is shown by the fact that while his basal metabolic rate is approximately 90 W, he can only develop about 50 W when working over long periods. When suitably stimulated, however, he can develop short surges of up to 250 W.

Problem 1.2

A footballer of mass 70 kg is running with the ball at a speed of about 5 ms^{-1}, i.e. he has potential energy in the form of glycogen and ATP and the more readily usable ATP is being converted into kinetic energy.

(*a*) Find his kinetic energy and the power developed when he is stopped within 1 sec in a tackle.

(*b*) What is his deceleration (assumed constant) during this time, i.e. his g value, and how does it compare with the 40g said to be imparted by Cassius Clay's left hand to a stationary non-elastic target?

1.8 Friction

When frictional forces operate on a body the system is non-conservative. For example consider a book being pushed along a table with constant velocity v (Fig. 1.14). The body is in fact in equilibrium under the action of 4 forces: the weight of the body $m\mathbf{g}$, the reaction force $\mathbf{R}$, the force from the hand $\mathbf{F}_p$, and the frictional force between the book and the table top $\mathbf{F}_r$. The equilibrium is described by two sets of equations

$$\mathbf{R} = m\mathbf{g} \text{ (action and reaction system)}$$

and

$$\mathbf{F}_r = \mathbf{F}_p$$

It is found experimentally that these two equations are further related by a *coefficient of friction* μ_f where

$$\mu_f = \frac{F_r}{R} \qquad 1.12$$

It is also found experimentally that μ_f is relatively independent of the velocity of the body (in this case the book) and of the area of the body in contact with the surface over which it is moving. μ_f depends in fact on the textures of the two surfaces.

Frictional forces also operate at the molecular level when molecules flow past one another. The coefficient of friction in fluids is called *viscosity* (Chapter 4, p. 50)

1.9 Circular Motion (an example of Simple Harmonic Motion)

Consider a body of mass m travelling round a circular path (Fig. 1.15a) of radius r, with a tangential velocity v. The time to go round the circle is called

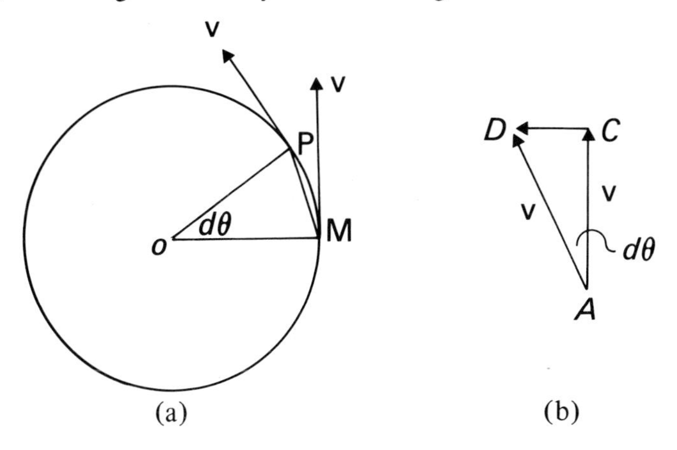

Fig. 1.15. (a) Circular motion: a body moves in a circle with tangential velocity v. (b) In the limiting case, when M and P are very close together, $CD = d\text{v}$ and $MP = rd\theta$. The direction of **CD** is towards the centre of the circle. (See text for fuller explanation.)

the *period* of the motion (T). As the length of circumference of the circle is $2\pi r$, then

$$v = \frac{2\pi r}{T} \tag{1.13}$$

The frequency f is the number of cycles completed per second (Hz in SI) and is the reciprocal of the period. Hence

$$f = \frac{v}{2\pi r} \tag{1.14}$$

The angular velocity (ω) is the number of radians swept out by the radius in 1 second, i.e.

$$\omega = d\theta / dt$$

and as 2π radians are swept out in T seconds,

$$\omega = \frac{2\pi}{T} = 2\pi f \tag{1.15}$$

Hence

$$\omega = v/r \tag{1.16}$$

As the velocity of the mass is constantly changing direction, the mass will experience accelerating forces. To work out the acceleration, consider the velocities at the two positions M and P close together on the circle (Fig. 1.15b). **AC** and **AD** represent the velocity vectors at these positions and the change in velocity is represented by the vector **CD**. From triangles OMP and ACD,

$$\frac{dv}{v} = \frac{r d\theta}{r} \tag{1.17}$$

Hence the acceleration of mass is given by

$$a = \frac{dv}{dt} = v d\theta / dt$$

$$= v\omega \text{ or } v^2/r \tag{1.18}$$

As the acceleration of the mass is directed towards the centre of the circle (Fig. 1.15b), the inertial or *centrifugal force* of magnitude mv^2/r will be directed *away from the centre.* As the effects of gravitational and inertial forces are indistinguishable, plant roots will grow away from the centre when the plant is set to grow on a rapidly rotating turn-table, just as they normally grow towards the centre of the earth. The angle at which they grow depends on the resultant of the gravitational and inertial forces (Fig. 1.16).

Molecules in a rapidly spinning centrifuge tube experience accelerations that are several times g and so sedimentation is speeded up.

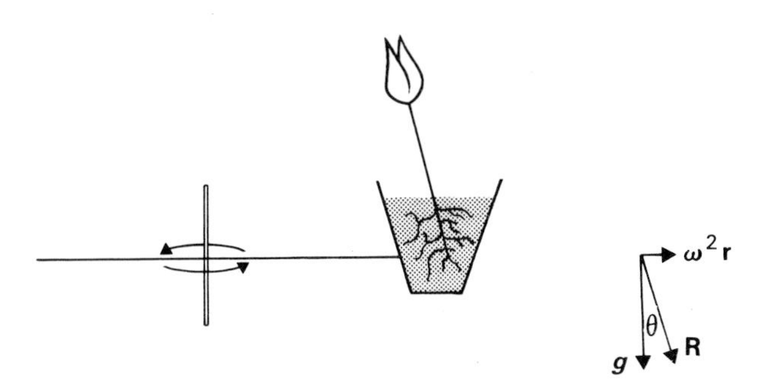

Fig. 1.16. If a flower-pot is attached to a horizontally rotating turn-table, the plant will grow at an angle θ to the vertical where θ is determined by the resultant **R** of the gravitational **g** and inertial (ω^2**r**) accelerations.

1.10 Rotating Vector Diagrams

Circular motion and wave motion (Chapters 6 and 7) are examples of *Simple Harmonic Motions* which can be represented by one or more rotating vectors. Light and sound waves can be represented in this way (Fig. 1.17). The magnitude of the vector represents the amplitude of the motion and the time taken for the vector to complete one revolution represents the period of the motion. The interaction of two

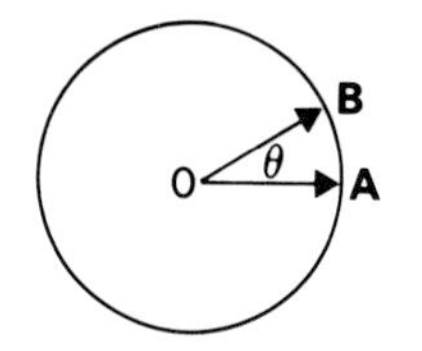

Fig. 1.17. Rotating vectors. The vector **OA** represents the motion at some arbitrary zero of time and **OB** represents the motion t seconds later. θ/t is the angular velocity of the motion and θ is the *phase* angle.

simple harmonic motions, e.g. the interaction of two sound (Chapter 6, p. 75) or light waves (Chapter 7, p. 96), can be represented by the resultant of the two vectors representing the individual motions.

1.11 Centre of Gravity

In the preceding sections it has been possible to discuss the physics of moving bodies by treating them simply as point masses. However, in many problems, e.g. in animal mechanics, both the size and shape of a body are important, and also the point of action of any forces on the body.

For example, the motion of a plank of wood depends on where it is pivoted. If it is a homogeneous plank then it will remain in equilibrium, i.e. in balance, if placed on a support at its centre. There are two forces on the plank, its weight mg acting vertically downwards, and a reaction force (Newton's third law) at the fulcrum acting upwards. At the point of balance, these forces are equal and opposite in magnitude and direction. This implies that there is a point in a body through which the weight of the body appears to act and this point is called the *centre of gravity* (Fig. 1.18).

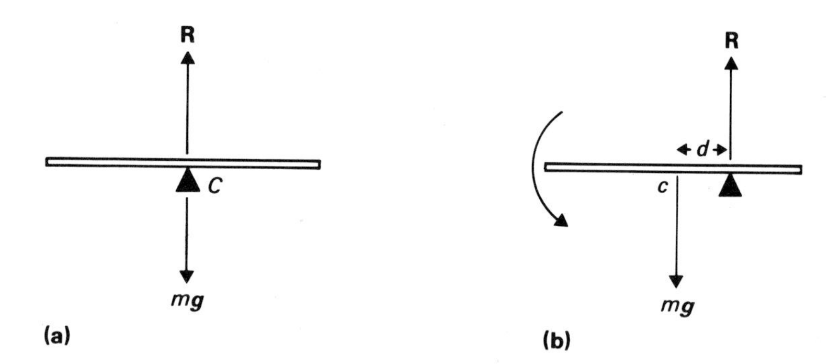

Fig. 1.18. (a) The homogeneous plank rests on a pivot at its centre and here it will be in equilibrium. (b) The fulcrum is towards the right and so the plank will swing downwards under the action of a force mg with moment mgd. C is the centre of gravity of the plank.

The centre of gravity of most bodies can be found simply by suspending the body from different points by means of a thread. Alexander (1968) has determined the centre of gravity of a locust by photographing a formalin-treated specimen. The centre of gravity is given by the intersection of the two lines AA' and BB' and the position in fact depends on the position of the locust's hind legs (Fig. 1.19).

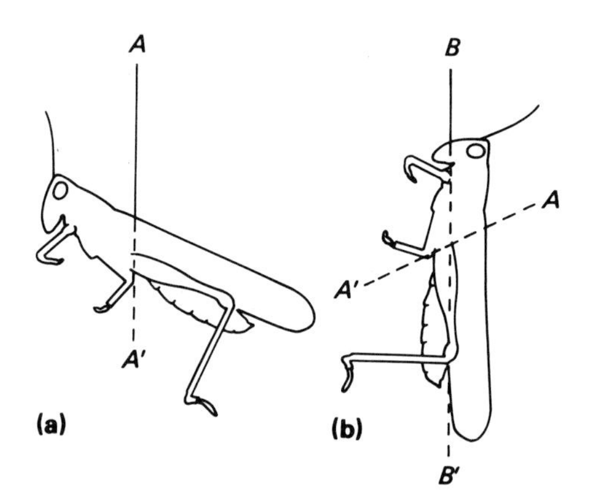

Fig. 1.19. Outlines traced from photographs of a preserved locust, suspended by a thread. The centre of gravity is at the intersection of the lines AA' and BB'. (From Alexander, 1968). Reproduced by permission of Sidgwick and Jackson Ltd.

1.12 The Lever

When a plank is pivoted near one end at P it can be used as a lever (Fig. 1.20). There are 4 forces acting on the plank: the downward force $\mathbf{F}_d$ used to try to lift the heavy mass M, the reaction force $\mathbf{R}$ at the fulcrum, and the forces arising from the gravitational pull on the plank and stone. Three of the four forces will tend to rotate the

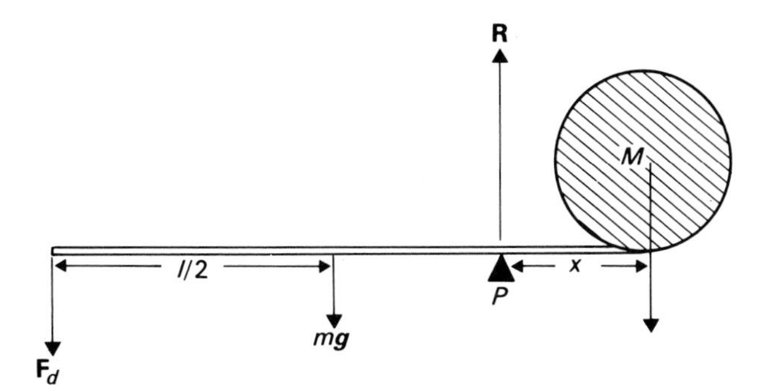

Fig. 1.20. A plank of mass m and length l pivoted near its end can be used to mechanical advantage. (See text for fuller explanation)

plank about the fulcrum and by how much depends on the product of the force and its perpendicular distance from the fulcrum. This product is called the *moment of the force* and the plank will be in equilibrium when the clockwise and counterclockwise moments are equal, i.e. when

$$F_d\,(l-x) + mg\,(l/2-x) = Mg\,x$$

$$F_d = \frac{Mgx + mg\,(x - l/2)}{l-x}$$

Hence F_d will be small if x is small and so the point of pivot should be as near the stone as possible. The ratio Mg/F_d is known as the *mechanical advantage* of the lever.

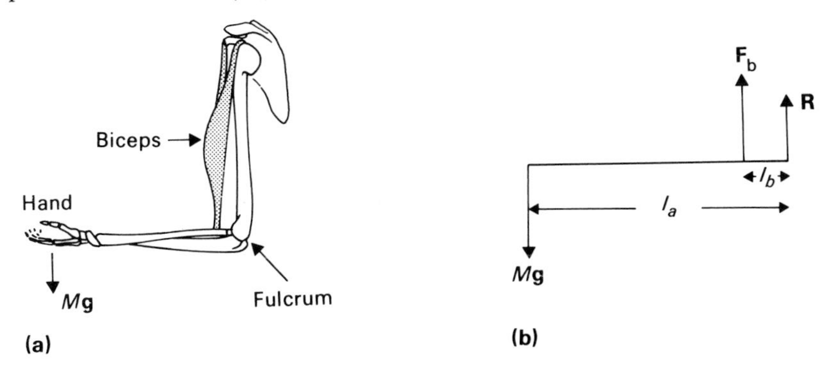

Fig. 1.21. Diagram of biceps muscles lifting a weight Mg in the hand. In this system the mechanical advantage l_b/l_a is less than 1.

When the bicep muscles of the arm are used to lift a weight Mg in the hand (Fig. 1.21) then the mechanical advantage of this lever system is given by

$$Mg/F_b = \frac{l_b}{l'_a}$$

Hence the force applied by the biceps is greater than the weight to be lifted; mechanical advantage < 1.

Nutcrackers and jaws are other examples of levers. Among mammals there are considerable differences in the shape of the lower jaw (lever) and the relative sizes of the various jaw muscles (forces). These are correlated with the very different ways in which mammals with different feeding habits use their jaws (Smith & Savage, 1959).

1.13 Motion of Rigid Bodies

The kinetic energy of a point mass moving with velocity v is $\frac{1}{2}mv^2$. To compute the kinetic energy of a moving rigid body, the shape of the body must be known, and also the axis about which the body is rotating. A simple example of the computation is provided by the insect's wing (Alexander, 1968). The wing is beating up and down about the axis indicated in Fig. 1.22.

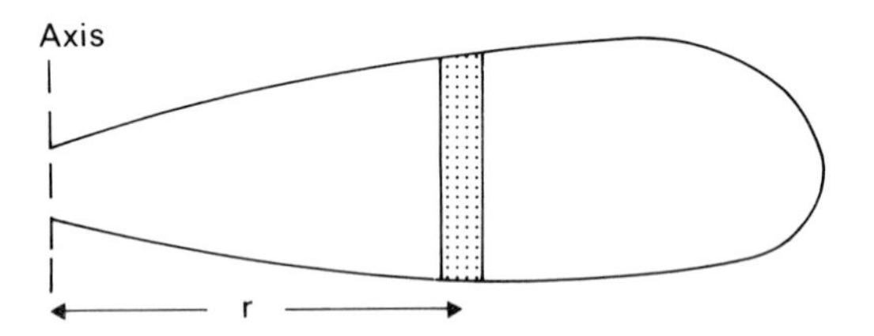

Fig. 1.22. A diagram of an insect wing illustrating the account of moments of inertia. (From Alexander, 1968)

Consider the wing to be divided into narrow strips cut parallel to the axis. The kinetic energy of a strip of mass m is given by

$$KE = \tfrac{1}{2} mv^2 = \tfrac{1}{2} mr^2 \omega^2 \qquad 1.20$$

where r is the distance of the centre of gravity of the strip from the axis and ω is the angular velocity of the strip about the axis. The kinetic energy of the whole wing will be given by the sum of the kinetic energies of all the strips, i.e.

$$\text{total } KE = \sum \tfrac{1}{2} mr^2 \omega^2$$

$$= \tfrac{1}{2} \omega^2 \sum mr^2 \qquad 1.21$$

Hence if each strip is carefully weighed and its distance from the axis of rotation also noted, the kinetic energy can be computed.

The sum Σmr^2 is called the *moment of inertia* of the wing and is denoted by I, i.e. the rotational kinetic energy of any body is given by $I\omega^2$. Sotavolta has used the I value for the insect's wing to compute its energy balance in flight (Alexander, 1968).

1.14 Elasticity: Hooke's Law

The important property of elastic materials is that they tend to return to their original shape when stretched. Elastic restoring forces are of great importance in biological systems as they provide a means of storing energy, e.g. kinetic energy imparted to the blood by the heart is stored in the elastic-walled arteries and serves to smooth the blood flow pattern.

Perfectly elastic bodies are said to obey Hooke's Law (Fig. 1.23) which states that the force $\mathbf{F}$ required to stretch the body is directly proportional to the extension x, i.e.

$$\mathbf{F} = kx \qquad 1.22$$

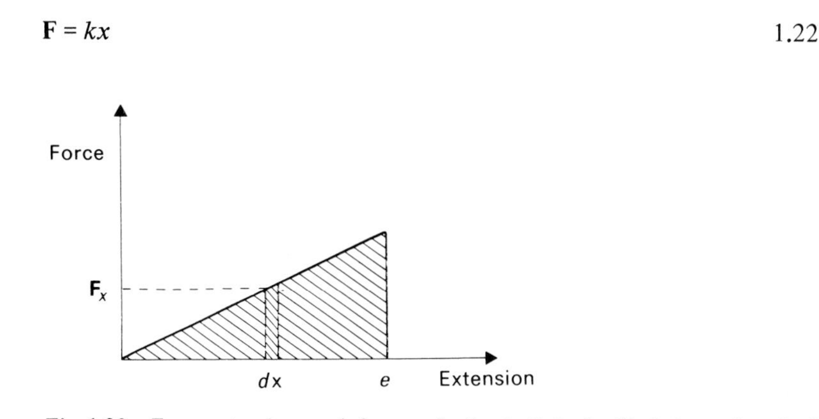

Fig. 1.23. Force extension graph for a perfectly elastic body. Work done when the force $\mathbf{F}_x$ moves its point of application through a small distance $d\mathbf{x}$ is $F_x dx$ and this is given by the area of the small strip. The total work done in extending by an amount e is the total area under the graph.

k is the stiffness constant and has units Nm^{-1}. The work done in extending the elastic body through a distance e is given by

$$\text{work} = \int_0^e F dx \qquad 1.23$$

$$= \int_0^e kx\, dx \qquad 1.24$$

$$= \tfrac{1}{2} ke^2 \qquad 1.25$$

and this is the *elastic potential energy* stored in the stretched body. It is therefore the work which the stretched body can do when released.

1.15 Volume Elasticity

The force per unit cross-sectional area on an elastic body fixed at one end is called the *tensile stress* (Fig. 1.24).

$$\text{tensile stress} = F/A \tag{1.26}$$

and has units $\mathrm{N\,m^{-2}}$.

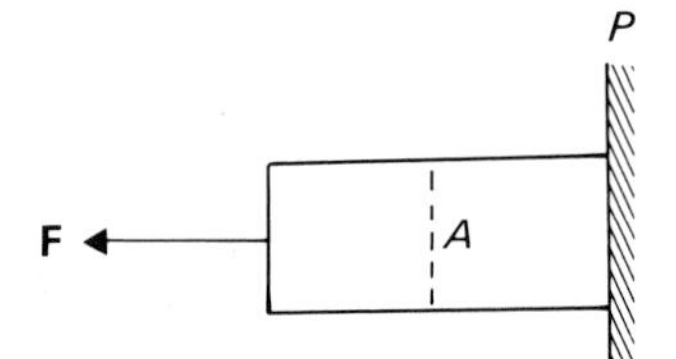

Fig. 1.24. An elastic body, attached at P and subject to a force **F**, experiences a tensile stress equal to F/A where A is the cross-sectional area of the body.

The term *strain* refers to the relative change in dimensions or shape of a body which is subject to stress (Fig. 1.25). The *tensile strain* is defined by the ratio e/l. which is dimensionless.

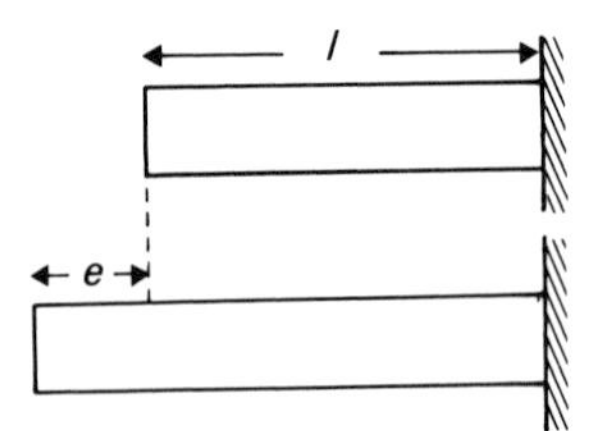

Fig. 1.25. An elastic body, originally length l is stretched by a small amount e, then the tensile strain is given by e/l.

When the force is parallel to the fixed edge (Fig. 1.26), the body undergoes *shear* deformation. The *shear stress* is the force divided by the area of the body and the *shear strain* is defined by the angle θ (radians).

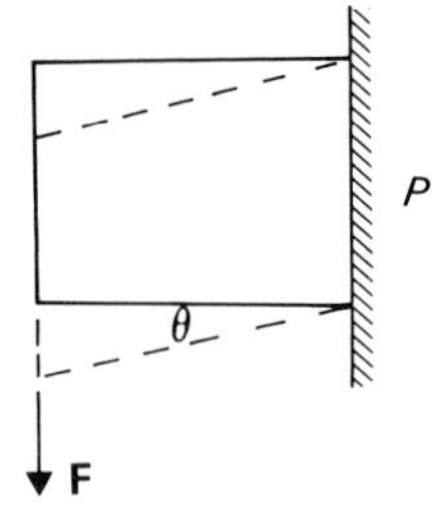

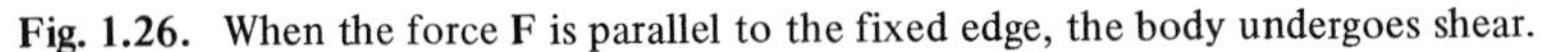

Fig. 1.26. When the force **F** is parallel to the fixed edge, the body undergoes shear.

1.16 Young's Modulus

For perfectly elastic bodies, the ratio of tensile stress to strain is a constant Y, and is another way of expressing Hooke's Law.

$$\text{Young's modulus} = \frac{\text{tensile stress}}{\text{tensile strain}} = \frac{F/A}{e/l} = Y \qquad 1.27$$

i.e.

$$F = \frac{YA}{l}e \qquad 1.28$$

and comparing equations 1.22 and 1.28

$$k = \frac{YA}{l} \qquad 1.29$$

Table 1.1. Although all materials are to some extent elastic, this table shows that materials commonly thought of as elastic, e.g. rubber, have a low modulus of elasticity, i.e. they undergo relatively large deformations when stressed.

Material	Young's Modulus (Nm^{-2})
Steel	2×10^{11}
Rubber	2×10^{6}
Resilin	1.7×10^{6}
Elastin	6×10^{5}

The stress-strain curve for a piece of rubber does not follow a straight line (Fig. 1.27) and nor do the stretching and relaxing paths coincide. Energy is in fact lost in the cycle due to the frictional interaction of the molecules comprising the stretched material, and the amount is the area contained in the so-called *hysteresis* curve. Materials

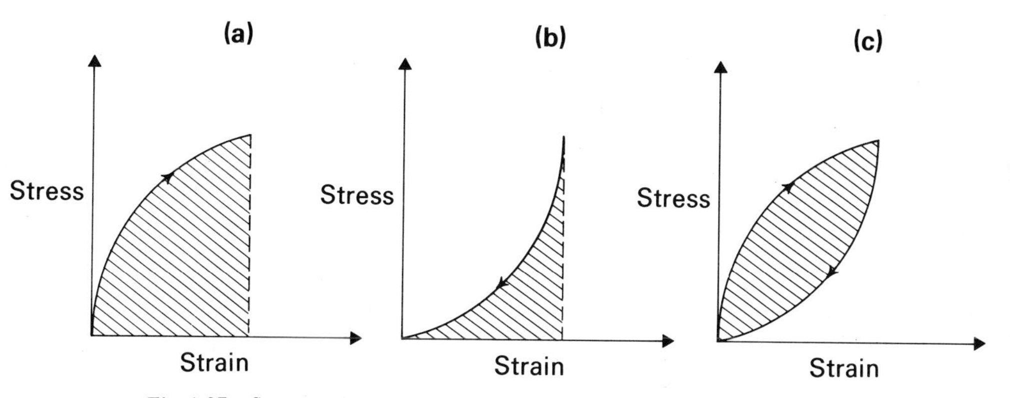

Fig. 1.27. Stress-strain curves for rubber. (a) The work done in stretching is the area under the curve. (b) The useful work recovered by the stretched body is released and is the area under the curve. (c) The total energy lost due to frictional interactions of the rubber molecule is given by the difference between the areas in (a) and (b). i.e. the hysteresis area.

with very large hysteresis curves are very useful as vibration absorbers. During each vibrational cycle, the applied mechanical energy is largely dissipated in the form of heat, and so is not passed on through the system.

The elastic material called resilin found in many arthropods and serving different functions is remarkable in that it absorbs less than 5% of the vibrational energy even when forced to vibrate at 200 cycles per second (Hz). It is found at the base of insect's wings and when the wing is raised during flight the resilin is compressed and the stored energy is released to facilitate the down stroke. The scallop *Pecten* makes use of an elastic protein called abductin while swimming. The large adductor muscle closes the shell and compresses the ligaments which are then ready to force the shell open when the muscle relaxes. The scallop swims by opening and closing the shell 3 times per second and at this frequency abductin has a very low hysteresis loss (Fig. 1.28).

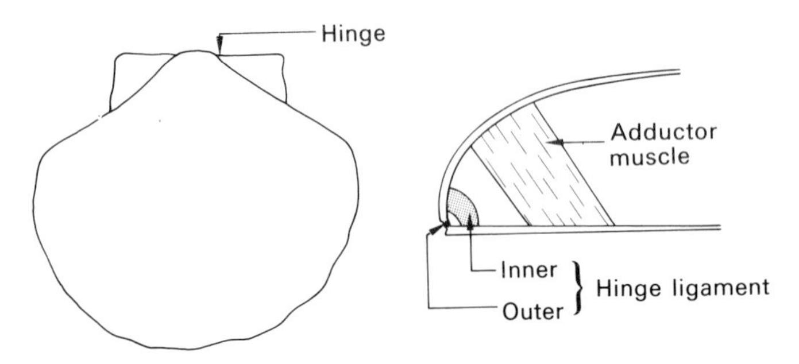

Fig. 1.28. Lateral view and diagrammatic transverse section of a scallop (*Pecten*), showing the position of elastic ligaments. (From Alexander, 1968). Reproduced by permission of Sidgwick and Jackson Ltd.

Problem 1.3 (From Bennett-Clark & Lucey, 1967)
Data collected from film records of the normal velocity of the ballistic part of the jump of a flea.

(*a*) The flea takes off with a vertical velocity of 1 $m s^{-1}$ and reaches a height of 3.5×10^{-2} m. Show that the deceleration due to air resistance is half the gravitational deceleration.

(*b*) The flea reached *from rest* its peak speed by 10^{-3} s. Show that the acceleration during this period was approximately 100 x gravity. (This compares with the acceleration of 5g on the body at lift-off in an Apollo moon-shot.)

(*c*) Given that the mass of the flea is 0.45×10^{-6} kg, find its kinetic energy after 10^{-3} s.

(*d*) The maximum output of insect muscle is known from other sources to be 60 W kg^{-1} muscle. Hence, assuming that 20% of the flea weight is muscle, show that muscle alone cannot power the flight.

(*e*) What is the energy store?

The answer lies in a resilin pad at the base of the hind leg. Bennett-Clark and Lucey noted that the flea slowly bent its hind legs prior to a jump. In this way it would compress the resilin pads and some unknown trip mechanism could release the energy for the jump. All that remains is to show that the 2 pads each of volume 1.4×10^{-4} mm^3 can store sufficient energy.

Assuming that resilin is perfectly elastic, the stored energy will be given by

$$PE = \tfrac{1}{2}\, ke^2 = \tfrac{1}{2}\, \frac{YA}{l} \,.\, e^2$$

The PE stored in 1 mm^3 which is compressed by 100% will be given by

$$PE = \tfrac{1}{2} \times 1.7 \times 10^6 \times 10^9 \quad \text{(using } Y \text{ value given in Table 1.1)}$$

$$= 0.9 \times 10^{-3} \text{ Jmm}^{-3}$$

(*Note*: this is only an order of magnitude calculation as the strain is not small.)

The total stored energy from 2.8×10^{-4} mm^3 of pad is therefore 2.5×10^{-7} J per flea, which is just sufficient to power the jump.

References

Alexander R.M. (1968) *Animal Mechanics*. Sedgwick and Jackson, London.

Alexander R.M. (1971) *Size and Shape*. Arnold, London.

Bennet-Clark H.C. & Lucey E.C.A. (1967) The Jump of the Flea; a Study of Energetics and a Model of the Mechanism. *J. Exp. Biol.* **47**, 59-76.

Ferrar W.L. (1967) *Calculus for Beginners*. Clarendon Press, Oxford.

Jarman M. (1970) *Examples in Quantitative Zoology*. Arnold, London.

Sears F.W. & Zemansky M.W. (1964) *University Physics*. Addison-Wesley, Reading, Mass.

Smith J.M. & Savage R.J.G. (1959) The Mechanics of Mammalian Jaws. *School Science Rev.* **40**, 289.

Chapter 2
Temperature and Heat

2.1 Temperature and Thermal Equilibrium

We have seen that in order to describe the motions of particles and rigid bodies, three fundamental quantities are required, namely mass, length, and time. These three quantities are sufficient to describe isolated bodies, but they have to be supplemented when bodies come into contact, and also when radiation falls on an isolated body. For example if we touch another body we say it feels either *hot* or *cold* relative to ourselves. In order to describe this feeling further, an additional fundamental quantity called *temperature* is invoked.

Two systems in thermal contact are said to be in *thermal equilibrium* when the temperatures of both bodies are equal. If two systems in contact are not in equilibrium, then heat energy will flow from the hot body to the cooler one until the temperatures are equal. A system can be insulated from those around it so that there is no thermal interaction. Such a system is said to be in *adiabatic isolation.*

Temperature measurements are made quantitative by the introduction of a temperature scale and the two systems most widely used are the Centigrade and Kelvin scales. In the former, the melting point of pure ice at one atmosphere pressure is taken as the arbitrary zero of temperature (0°C) while the boiling point of pure water at one atmosphere pressure is taken as 100°C. In the latter scale the melting point of ice is 273 degrees Kelvin (273°K) and the boiling point of water is 373°K.

As well as being defined in terms of the thermal equilibrium of bodies, temperature can also be defined in terms of the kinetic energy content of the molecules comprising the body (internal energy of the body). The kinetic energy of a gas molecule, for example, depends solely on the temperature of the gas and from the Kinetic Theory of Gasses (Sears & Zemansky, 1964) the average kinetic energy per molecule is given by

$$\mathrm{KE} = \tfrac{3}{2}\, kT \qquad 2.1$$

where T is the temperature of the gas in °K and k (the Boltzmann constant) is equal to 1.38×10^{-23} J°K^{-1}. The absolute zero of temperature 0°K is reached when the molecular kinetic energy is zero.

A *thermometer* is a device for measuring temperature and the most common type, a liquid in glass thermometer, makes use of the fact that liquids expand when heated.

Thermocouples (Fig. 2.1) are widely used for temperature measurements in biological systems and in these a thermoelectric voltage, usually measured with a galvanometer, is produced at the junction of two dissimilar wires. The usefulness of the thermocouple arises because the area of the junction in contact with the system under investigation can be quite small and also because it is simple to get relatively accurate measurements.

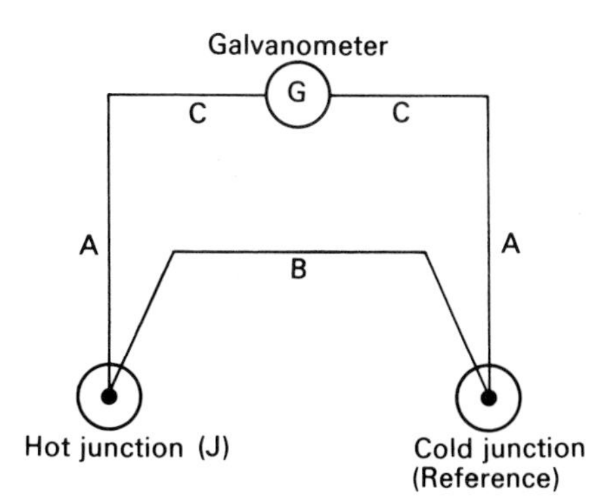

Fig. 2.1. An EMF (Chapter 9) is developed between the hot and cold junctions of two dissimilar metals A and B. These are connected, via copper wires C, to a simple galvonometer G, and the deflection observed can readily be calibrated in terms of the temperature difference between J and the cold bath which is commonly held at 0°C by stirred, melting ice.

2.2 Heat and the First Law of Thermodynamics

Joule was the first to demonstrate two ways of increasing the temperature of a body in a quantitative manner. He showed that the final temperature of a can of water does not depend on the manner in which the energy is delivered. Y units of heat energy from a bunsen flame produce the same temperature rise as Y units of mechanical energy dissipated during vigorous stirring.

Joule's conclusion is stated in thermodynamic terminology by the following equation, called the First Law of Thermodynamics

$$\Delta E = \Delta Q + \Delta W \tag{2.2}$$

where ΔE is the increase in internal energy of the water, ΔQ is the heat energy supplied to the water, and ΔW is the work done on the water. The temperature of the water rises because there has been an increase in internal (thermal) energy.

2.3 Heat Capacity

When a body receives an amount of heat ΔQ then its temperature rises. The extent of the temperature change (ΔT) depends on the mass of the body m and its specific heat S.

The *specific heat* of a body is defined as the heat required to raise the temperature of a unit mass of the body through 1 degree (1°C or 1°K). The specific heat S is therefore defined by the equation

$$\Delta Q = mS\,\Delta T \qquad 2.3$$

Examples of the specific heat of some common substances are given in Table 2.1.

Table 2.1 Specific Heat

Substance	Specific Heat (J kg^{-1} deg^{-1})
Copper	390
Ice	2110
Water	4200
Biological Tissue	about 3700
Air	950 (STP)

The specific heat for air is given at 0°C and at a constant pressure of 1 atmosphere (10^5 Nm^{-2}) i.e. the specific heat is said to be quoted at STP (Standard Temperature and Pressure).

2.4 Heat and Change of Phase

All substances exist in one of three phases – solid, liquid, and gas and the change from one phase to another is marked by a change in the internal energy of the substance. For example, one kg of water vapour has a much higher internal energy than one kg of water at the same temperature so that a relatively large amount of heat is required to vaporize the water. This heat is called the *latent heat of vaporization* and for water it is approximately 2.5×10^6 J kg^{-1}.

Warm blooded animals make use of latent heat as one means of regulating their body temperature. When the hypothalmus detects an increase in blood temperature it signals the sweat glands to increase production. The vaporization of this sweat requires latent heat energy and this is supplied by the body which is therefore cooled and so the hypothalamus activity is depressed. This is a simple example of a biological feedback control mechanism.

2.5 Thermal Interaction with the Environment

All bodies, whether located for example on a sunny Norfolk beach or in a spacecraft in the vastness of space, exchange heat energy with their environment. If we consider for a moment a body on the beach, we see that it will be heated by energy radiated from the sun above and energy conducted from the sand beneath. It will probably also be cooled by winds (convected energy) from the sea. If the body is alive, then it too will be producing heat energy from chemical reactions. As all living systems can only work within a small temperature range, special devices (fur, sweating, large ears, etc.) are developed in order to maintain a stable temperature. We shall now consider the mathematics involved in describing heat transfer.

2.6 Heat Transfer by Conduction

When a poker is held with its tip in glowing coals, then the handle becomes gradually hotter although it is not in direct contact with the heat source. There is no gross motion of the body during this heat transfer, which is said to take place by conduction.

It is an experimental fact that the rate of heat flow depends on two factors. Firstly it is proportional to the cross-sectional area through which the transfer is taking place, and secondly it depends on the temperature gradient between the source and receiving parts of the body.

If we consider a sheet of material (Fig. 2.2) of cross-sectional area A and thickness ΔX, with face 1 maintained at a temperature T_1 and face 2 at a temperature T_2, then heat will flow from 1 to 2 when $T_1 > T_2$. i.e. thermal energy travels down a temperature gradient.

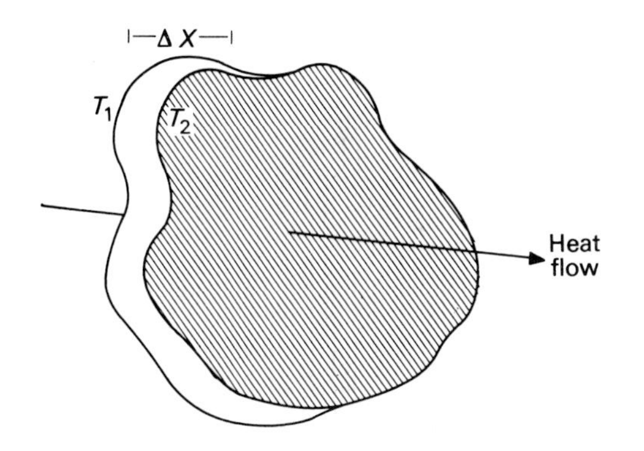

Fig. 2.2. Section of a sheet of material of thickness ΔX and cross-sectional area A. The temperatures of the two faces are T_1 and T_2 respectively and when $T_1 > T_2$ then heat flows from face 1 to 2. The rate of heat transfer is given by Fourier's equation.

This is expressed mathematically by Fourier's equation

$$H = KA\frac{\Delta T}{\Delta X} \qquad 2.4$$

where H is the heat flux in watts, K is the thermal conductivity of the material in $\text{Wm}^{-1}\ \text{deg}^{-1}$ and $\Delta T/\Delta X$ is the temperature gradient.

Equation 2.4 has the form

flow = constant x driving force

and in this case the driving force on the heat flow is the temperature gradient. A similar equation can be written for the flow electric current I through a conductor, driven by a voltage difference V across the ends of the conductor (Chapter 9, p. 148).

$$I = \text{constant} \times V$$

and in this case the constant is the electrical conductance which is the reciprocal of resistance.

Heat transfer in good conductors is in fact effected by the flow of electrons and good conductors of heat also conduct electricity well. In insulators, heat flow takes place by the vibrations of the constituent molecules, a much less efficient process. Crystalline substances are good conductors of heat and electricity and this explains why ice is a better conductor than water (Table 2.2).

Table 2.2. Thermal Conductivity K. The units of K are $Wm^{-1}\ deg^{-1}$.

Substance	K	Substance	K
Air (at STP)	0.024	Glass	1.0
Water	0.59	Ice	2.1
Brick	0.6	Silver	418

As an example in heat conduction, consider the heat loss from a small room at 25°C to the outside at 0°C through a window of area $2m^2$. The window thickness is 4 mm. Then assuming all the heat loss takes place via the window it is given by

$$H = KA\ \Delta T/\Delta X$$

$$= 1.0 \times 2 \times 25/4 \times 10^{-3}$$

$$= 12.5 \times 10^3 \text{ or } 12.5 \text{ kW}$$

From this prohibitively massive heat loss, it is obvious that some vital factor has been omitted from the calculations. This factor is the so-called stagnant or unstirred layers of air that surround every body. It is very difficult to estimate the magnitude of these, but suppose they are only 2mm thick on either side of the window, then the heat loss will be cut to approximately 0.2 kW. (Compare K for air and glass from Table 2.2.) The above calculation shows the importance of the trapped air space in double-glazed windows.

When the stagnant layers are disturbed, e.g. by a wind blowing on the external surface of the window, then the heat loss is much greater and in fact takes place by *forced convection.*

You can have a personal experience of the insulating power of an unstirred layer in a sauna bath where the air temperature is over 100°C, far above that required to produce severe burns. If you blow on your arm and disturb the layers then you will feel a burning sensation. These layers will also help retain heat when you emerge to the snow outside. If, however, the full therapeutic value of a sudden temperature shock is required, they can be disturbed by a suitable application of birch twigs.

2.7 Heat Transfer by Convection

Convection is a means of heat transfer in fluids and takes place through a movement of the fluid itself.

An example of convection and one which shows the simple principles of central-heating systems is provided by a U-tube filled with water (Fig. 2.3a). One arm of the U-tube carries an open extension reaching the other arm. When water is first poured

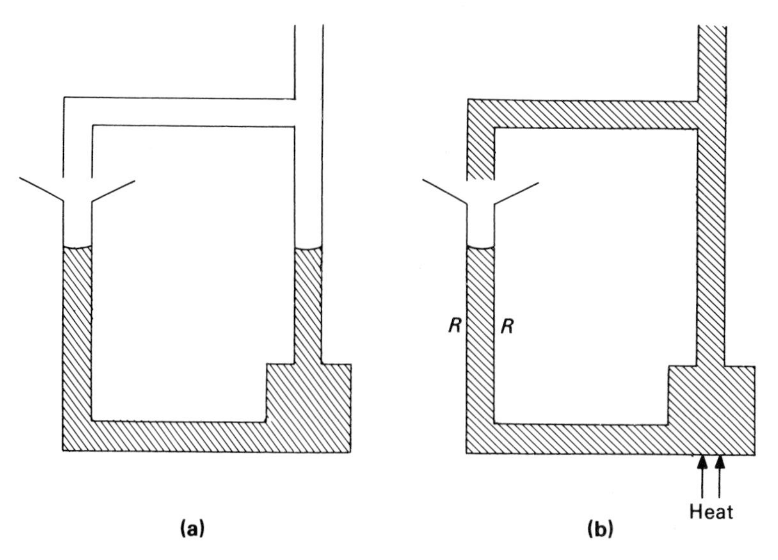

Fig. 2.3. When heat is supplied to the system (b) the water in the right-hand column expands and water circulates round the system.

into the U-tube, the same level is of course reached in both arms. If the base of the arm carrying the extension is heated, then the water at the base will expand and so become less dense. It will therefore rise, carrying its heat to the water molecules above it. When the water in the right hand arm rises to the level of the extension it will pour over into the left hand column and provided heat is removed from this column the water will continue to circulate (Fig 2.3b). The cold side (R) in fact corresponds to the radiators of a heating system and the hot side to the boiler. For maximally efficient heat transfer all modern central heating systems also pump the water round giving forced convection.

The fact that water is most dense at 4°C, and not at 0°C has important consequences for all water-dependent ecosystems in winter. Starting at a temperature above 4°C, then as the surface of a pond is cooled by a drop in air temperature, convection currents will carry the colder denser water down until the water temperature is 4°C when convection stops. From then on the colder, *lighter* water will remain at the surface, convection will cease and heat will only be lost by conduction. As water is a poor conductor, this will be a relatively slow process. If further cooling occurs, then ice will form and as ice is lighter than water at 0° C, it stays on the surface and, despite its greater conductivity, helps to reduce heat loss by preventing wind from churning up the pond and bringing warmer water to the surface.

2.8 Forced Convection

When a pump forces water round a heating system, or a fan blows on the surface of a hot body, then the heat transfer is by forced convection. Newton's law of cooling describes thé rate of heat loss H from a body of unit surface area in a cool air stream, for example

$$H = K_c (T_1 - T_2) \quad 2.5$$

when T_1 is the temperature of the body, T_2 the air temperature and K_c is the convection coefficient, with units Wm^{-2} deg^{-1}.

Let us consider for a moment the processess involved in heat transfer by forced convection, e.g. at the surface of a leaf. The transfer will depend on whether or not the flow is laminar (Chapter 4, p. 50) across the leaf surface; hence from equation 4.8 it will depend on the density, viscosity, and velocity of the air. Heat will be removed from the surface of the leaf by conduction and it is then removed from the vicinity by convection. K_c depends therefore on how far laminar flow extends outwards from the surface and it depends on the heat capacity and thermal conductivity of the cooling fluid stream. For a flat plate in air (Gates, 1965).

$$K_c = 4\sqrt{\frac{\mathrm{v}}{L}}\ \mathrm{Wm^{-2}\ deg^{-1}}\ * \quad 2.6$$

where v is the stream velocity and L is the length dimension of the plate. This means that the smaller the surface dimension along the direction of flow the greater will be the heat loss, because heat transfer takes place mainly at the edge of the plate. This is in fact one reason why leaves are divided into lobes as it enables them to have an efficient heat loss.

It will be shown in Problem 2.2 that forced convection is probably the most efficient way of removing excess heat from a leaf surface.

2.9 Heat Transfer by Radiation

Two bodies at different temperatures will exchange heat even when there is no pos-

* This equation is dimensionally inhomogeneous and in fact the 4 here has units. (See Gates, 1965.)

sibility of exchange by conduction or convection; the transfer of heat takes place in fact by *radiation*. Good radiators of heat are also good absorbers and the most efficient absorber possible, called a *black body*, is one which will absorb all the radiation falling on it. Such a body also reradiates all the energy falling on it. The human skin, white or black, is a good approximation of a black body.

The total emissive power of a body ϵ_0 is defined as the total radiant energy of all wavelengths emitted by the body per square metre of its surface per second.

For a black body the total emissive power is proportional to the fourth power of the temperature (Kelvin), i.e.

$$\epsilon_0 = \sigma T^4 \qquad 2.7(a)$$

This is the *Stefan-Boltzmann Law* and the universal constant σ, called Stefan's constant, has the value 5.69×10^{-8} Wm^{-2} $°K^{-4}$. The total emissive power of any other body ϵ is a fraction of this and this fraction is called the *emissivity* (e) of the body, i.e.

$$\epsilon = e\,\epsilon_0 \qquad 2.7(b)$$

If a beam of radiant energy from a black body is dispersed into a spectrum, and if the energy content at any wavelength is measured, then a maximum in energy is found and this maximum is dependent on the temperature of the radiating body (Fig. 2.4a and b).

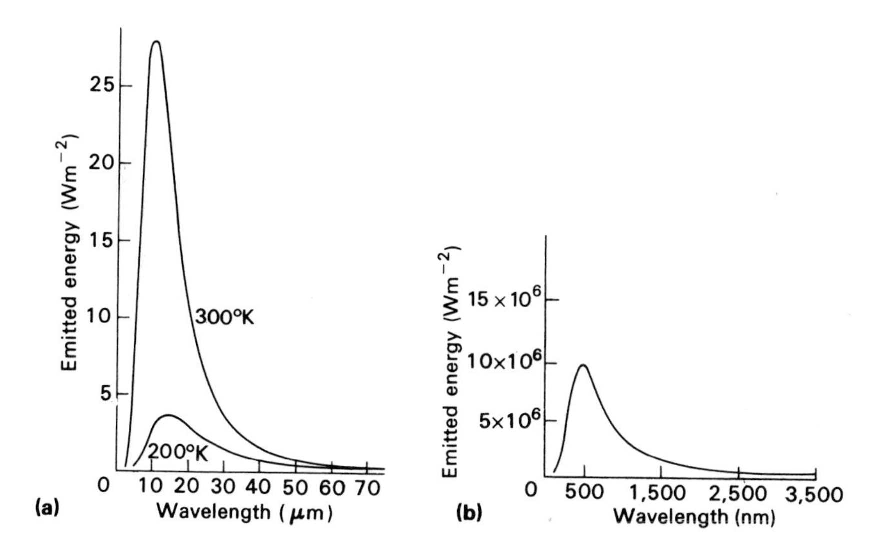

Fig. 2.4. Black-body emission curves for (a) two bodies with temperatures of 300° K and 200° K, and (b) the sun at 6,000° K. Note the contrasting scales of emitted energy in units of Wm^{-2} and the contrasting wavelength scales between (a) and (b). (From Barry & Chorley, 1968)

The energy emitted from unit area of the emitting surface *in unit time* within a small wavelength range $d\lambda$, centred on the wavelength λ, is $E_\lambda \, d\lambda$ (Sears & Zemansky, 1964)

where $$E_\lambda \, d\lambda = \frac{C_1}{\lambda^5 (e^{C_2/\lambda T} - 1)} \, . \, d\lambda \text{ (Planck's formula)} \qquad 2.8$$

$E_\lambda \, d\lambda$ has units Wm^{-2}, T is the absolute temperature, and C_2 is the constant 1.44 x 10^{-2} m°K. C_1 is another constant whose value is 3.74 x 10^{-16} Wm^2.

The value of λ at which the energy emitted is a maximum is approximately given by

$$\lambda_{max} = \frac{0.288}{T} \times 10^{-2} \text{ m} \qquad 2.9$$

This shift in relative intensities accounts for the change in colour of a body emitting visible light as its temperature is raised. At 2000°K a body is red hot and at 6000°K, a body is white hot. Above 10,000°K blue light is emitted with a greater intensity than red and a body is blue hot – some stars show this effect.

Problem 2.1

A recent report on food preservation in Britain (Guardian, 2nd February 1973) revealed that the temperature of meat in transparent plastic packs could be as much as 12°C above the freezer temperature when they were stored in a situation where they were exposed to illumination.

(*a*) Explain how this high temperature effect occurs.

(*b*) Design a pack which would prevent this.

Problem 2.2

The total energy falling on a horizontal leaf from the sun and the immediate environment is 800 Wm^{-2}.

(*a*) If we assume to begin with that no heat is lost or reflected from the leaf, what would be the rate of rise of the leaf's temperature, given that the specific heat and weight per unit area of the leaf are 3.8 x 10^3 J kg^{-1} °K^{-1} and 0.1 kg m^{-2}?

(*b*) Think for a moment about the possible ways in which the leaf might get rid of this unwanted energy.

(*c*) Calculate the temperature the leaf would reach if it lost all its heat by radiation.

(*d*) As you will now have shown that the temperature of the leaf is still too high, calculate the transpiration rate necessary to maintain the temperature at the reasonable level of 40°C.

(*e*) If a wind velocity 4.5 ms^{-1} (10 m.p.h. for the conservatives) is blowing across the leaf, calculate the fraction of the total incident radiation that would be lost by forced convection when the temperature difference between the lead and its environment is (i) 5°C and (ii) 10°C. Assume the leaf to be a square plate of linear dimension 0.01 m.

Answer

(*a*) Substitute the values given above in equation 2.3, hence

$$\frac{\Delta T}{\Delta t} = \frac{800}{0.1 \times 3.8 \times 10^3} \text{ deg s}^{-1}$$

$$= 2.1 \text{ deg s}^{-1}$$

which is a very fast temperature rise indeed.

(*c*) If we assume that heat is lost by radiation alone, then thermal equilibrium will be reached when this balances the incoming radiation, i.e. when

$$\sigma T^4 = 800 \text{ Wm}^{-2}$$

or

$$T = \sqrt[4]{\frac{800}{5.7 \times 10^{-8}}} \ {}^\circ\text{K}$$

$$= 344^\circ\text{K or } 71^\circ\text{C}$$

This temperature is greater than the highest recorded from a leaf (49°C) and so there must be some additional cooling process.

(*d*) Remember that the leaf will be reradiating some of the incoming energy, so the energy which has to be lost by transpiration will be the difference between the two.

Loss due to reradiation at 40°C (313°K) $= 5.7 \times 10^{-8} \times (313)^4 \text{ Wm}^{-2}$

$= 550 \text{ Wm}^{-2}$

Hence the amount to be lost by transpiration

$= 250 \text{ Wm}^{-2}$

Now as 2.5×10^6 J are required to vaporize 1 kg of water at 40°C, the excess energy will vaporize 10^{-4} kg m^{-2} s^{-1}. This is the required transpiration rate and it is a relatively high one. There is, however, another possible method for heat loss.

(*e*) The heat loss carried by forced convection is given by

$$\Delta Q = K_c \, \Delta T \qquad \text{(equation 2.6)}$$

$$= 424 \text{ Wm}^{-2} \text{ when } \Delta T = 5^\circ\text{C}$$

and

$$= 848 \text{ Wm}^{-2} \text{ when } \Delta T = 10^\circ\text{C}$$

Now you should have sufficient insight to investigate temperature regulation in other systems. An interesting starting point is the question of heat loss and body temperature in flying insects studied by Church (1960).

References

Barry R.G. & Chorley R.J. (1968) *Atmosphere, Weather and Climate.* Methuen, London.
Burns D.M. & MacDonald S.G.G. (1970) *Physics for Biology and PreMedical Students.* Addison-Wesley, London.
Church N.S. (1960) Heat Loss and the Body Temperatures of Flying Insects. *J. Exp. Biol* **37**, 171-185 and 186-212.
Gates D.M. (1965) *Energy Exchange in the Biosphere.* Harper & Row, London & New York.
Jarman M. (1970) *Examples in Quantitative Zoology.* Arnold, London.
Monteith J.L. (1973) *Principles of Environmental Physics.* Arnold, London.
Phillipson J. (1966) *Ecological Energetics.* Arnold, London.
Sears F.W. & Zemansky M.W. (1964) *University Physics.* Addison-Wesley, Reading, Mass.

Chapter 3
Fluids: Pressure and Gases

3.1 Introduction

We live in a fluid environment; the air flows round us, conveying the all-important oxygen to our blood, which in turn carries it to our tissues. Even the seemingly solid earth beneath us is in a constant state of flux. All *flows* of materials are set up by *driving forces* and we shall spend a large fraction of the following two chapters investigating the relationships between the forces and their induced flows. Firstly, however, we shall discuss the static properties of fluids and the relationships between pressures and volumes.

3.2 Pressure

A fluid in a container exerts a force on every part of the container it is in contact with, because every molecule of the fluid is in a continual state of motion. When a fluid molecule bumps against the container walls, its velocity is altered, and so it must exert some force. The fluid in fact exerts a *pressure.*

The pressure at any point is defined as the force per unit area surrounding the point.

$$\text{pressure} = \text{force/area} \qquad 3.1$$

and has units $\mathrm{N\,m^{-2}}$.

One common way of measuring air pressure is by means of the mercury barometer. When a glass tube is filled with mercury and inverted over a mercury bath (Fig. 3.1a), the mercury level falls until the column is in equilibrium with the air pressure. Considering an imaginary cut through B, parallel to the surface of the bath, we can then see that as mercury is incompressible, the pressure exerted on the upper surface of the cut by the lower is the atmospheric pressure, P. If the cross-sectional area of the cut is A, then this upward force will be PA. The downward force due to the mass of mercury above the cut is mg and at equilibrium

$$\begin{aligned} PA &= mg \\ &= \rho Agh \end{aligned} \qquad 3.2$$

where ρ is the density of the mercury and h the height of the column. Hence

$$P = \rho g h \qquad 3.3$$

At sea level, the average height which the air pressure can support is 0.76 m of mercury, and as the density of mercury is 13.6×10^3 kg m^{-3}, the atmospheric pressure is approximately 10^5 Nm^{-2}. In older texts, pressures are often given in terms of atmospheres or metres of mercury.

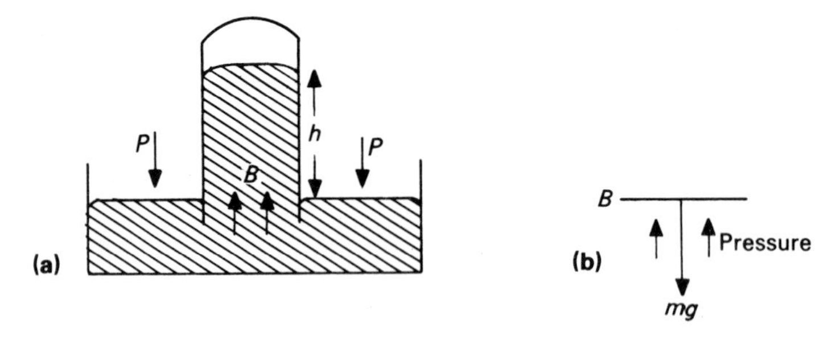

Fig. 3.1. (a) Air pressure can support a column of mercury of length h in an enclosed tube. (b) At the surface through B the atmospheric pressure P acting on the area A balances the weight of the column above it.

Problem 3.1

The mean arterial pressure at the heart level in man is often quoted as 100 mm Hg.

(*a*) Convert this into SI units.

(*b*) What is the pressure (i) at a point 0.5 m above the heart and (ii) at the feet, some 1.5 m below the heart. (Density of blood can be taken as 10^3 kgm^{-3}.)

The pressure P exerted by a gas depends on the volume V and temperature T of the gas, and on the number n of moles present. The relationship between the quantities can be measured experimentally. If the temperature is fixed during the experiment, and if the relationship PV/nT is plotted against pressure, then a smooth curve T_1, is obtained (Fig. 3.2). When the process is repeated at other temperatures T_2 and T_3 it is found that all the *isotherms* intersect the vertical axis at the same point.

It is further found that this point is the same for all gases. The point of intersection is called the *Universal Gas Constant* R and has the value

$$R = 8.3 \text{ J mol}^{-1} \text{ deg}^{-1}$$

It is customary to define an ideal gas as one for which PV/nT is equal to R at all pressures. The relationship

$$PV/nT = R \qquad 3.4$$

is also called the equation of state for an ideal gas.

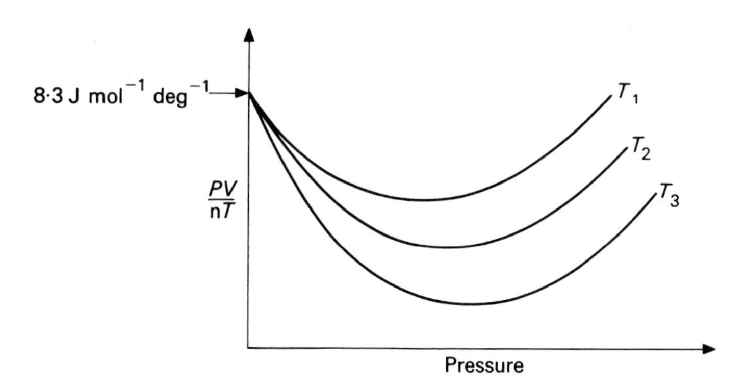

Fig. 3.2. The isotherms for a fixed mass of gas intersect the vertical axis at the same point. At low pressures, therefore, PV/nT = constant is true of all gases.

If we have a mixture of ideal gases, then the total number of moles n is given by

$$n = n_1 + n_2 + n_3 \qquad 3.5$$

and as $P = \dfrac{nRT}{V}$

$$P = \frac{n_1RT}{V} + \frac{n_2RT}{V} + \frac{n_3RT}{V}$$

$$= P_1 + P_2 + P_3 \qquad 3.6$$

$$= \sum_i P_i$$

Where P_i, called the *partial pressure,* is the pressure which the gas i would exert were it alone in the volume V. Equation 3.6 is called Dalton's Law of Partial Pressures. From the ideal gas equation, for a fixed mass of gas nR is constant and so PV/T is constant, i.e.

$$\frac{P_1V_1}{T_1} = \frac{P_2V_2}{T_2} \qquad 3.7$$

If the temperatures T_1 and T_2 are also the same, then

$$PV = \text{constant} \qquad 3.7a$$

This equation was discovered experimentally by Boyle in 1660.

3.3 Properties of Real Gases

According to Boyle's Law (equation 3.7a), a graph of pressure against volume should be a rectangular hyperbola, and there should be one smooth curve for each temperature. The isotherms of a real gas are shown below (Fig. 3.3). The upper curve A corresponds to the ideal gas behaviour and the substance remains a gas at all volumes. Curves B and C show a marked deviation from the ideal, but the substance still remains a gas.

For curve D, however, the situation is different. If originally the gas is under only a low pressure and this is then gradually increased, while the temperature is held constant, then a point a is reached when the volume decreases without any further increase to pressure. The gas begins to liquefy until finally a point b is reached when liquefaction is complete. Thereafter the isotherm rises steeply as liquids are virtually incompressible. Isotherm C is called the critical isotherm.

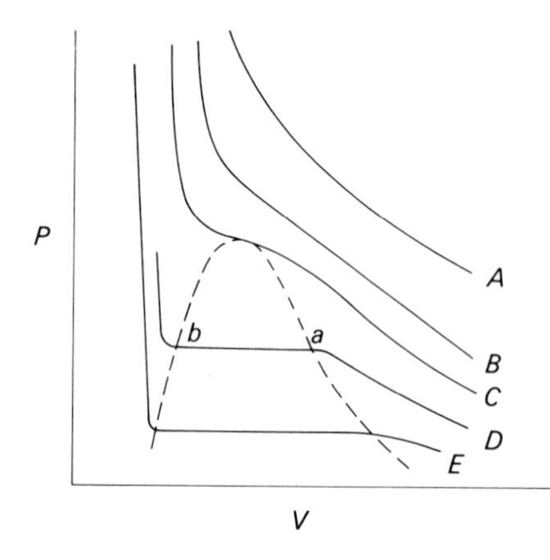

Fig. 3.3. Isotherms of a real gas deviate markedly from Boyle's Law because liquefaction occurs. Isotherm C is the critical isotherm.

The dotted curve encloses the region where the gas exists in equilibrium with its liquid. A gas below its critical temperature is referred to as *vapour*, and the pressure corresponding to the line ab is the *saturated vapour pressure* at that temperature. The saturated vapour pressure is the maximum pressure that can be exerted by the vapour at that temperature and is dependent only on the temperature (Table 3.1).

Table 3.1 Vapour pressure of water at various temperatures.

Temperature (°C)	Vapour pressure ($kN\,m^{-2}$)
0	0.60
5	0.87
10	1.21
15	1.69
20	2.30
25	3.14
30	4.18
35	5.55
40	7.26

The partial pressure of water vapour in the atmosphere at any temperature is usually less than the saturated vapour pressure for the same temperature and the ratio of the two expressed as a percentage is known as the *relative humidity,* i.e.

$$\text{relative humidity} = 100 \times \frac{\text{partial pressure of vapour}}{\text{saturated vapour pressure at same temperature}}$$

$$= 100 \times \frac{\text{amount of vapour the air contains}}{\text{amount it would contain if saturated}}$$

Relative humidity values are meaningless by themselves, and some other information, e.g. air temperature or one of the actual vapour pressures, is required.

Saturation can be achieved in two ways.

(i) The total water content can be increased until the pressure of the vapour is the saturated vapour pressure.

(ii) The temperature can be lowered until the actual amount of water vapour in the air is enough to cause saturation at the new temperature. It is this process that causes mist, fog, etc. The temperature at which moist air would be saturated is called the *dew point* and this constitutes an easily measured parameter. All that is necessary is to cool a brightly polished metal surface and observe the temperature at which it becomes clouded with moisture. The relative humidity is then simply the saturated vapour pressure at the dew point divided by the saturated vapour pressure at the actual air temperature times 100. For example, suppose the dew point measured in this way is 10°C when the air temperature is 20°C. We then know that the vapour in the air is saturated at 10°C, hence its partial pressure is 1.2 $kN m^{-2}$ equal to the saturated vapour pressure at 10°C. As the pressure necessary for saturation at 20°C is 2.3 $kN m^{-2}$,

$$\text{the relative humidity} = \frac{1.2}{2.3} \times 100 = 52\%$$

3.4 Swim Bladder

The ideal gas laws can be invoked to explain the mechanism of the swim bladder, which is a buoyancy device enabling fish to hover in mid-water with a minimum expenditure of energy. The swim bladder is a gas-filled organ within the body cavity of fish which reduces the overall density of the fish from about 1.07×10^3 $kg m^{-3}$ to near 1.0×10^3 $kg m^{-3}$.

To a certain extent the density of a fish is under its own control as some fish can secrete gas in and out of the swim bladder, while others have a special sphincter mechanism which allows for a rapid discharge of excess gas through the mouth.

For any fixed mass of gas, however, there will only be one depth at which the fish will be in equilibrium, and even this equilibrium is unstable as a slight upward movement will reduce the pressure on the gas, allow it to expand and so reduce the overall density of the fish, forcing it up still further. Hence, although the swim bladder aids the fish, a continuous expenditure of energy is required to maintain equilibrium.

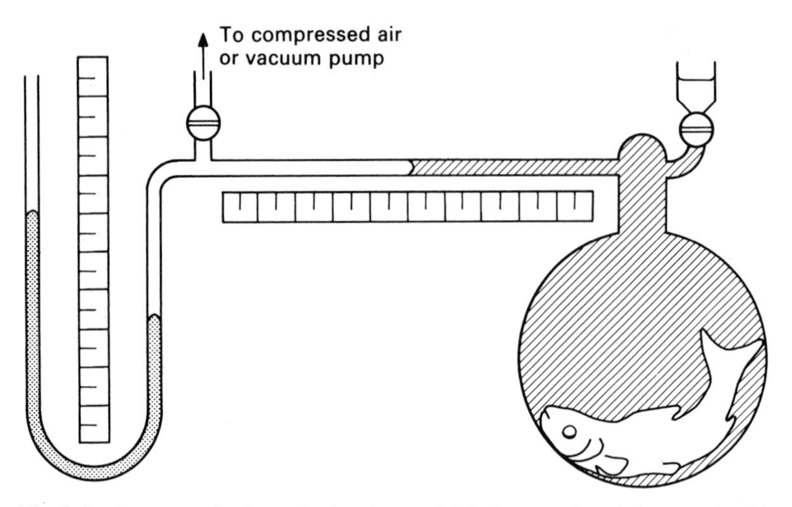

Fig. 3.4. Apparatus for investigating the mechanical properties of the swim bladders of fish. (From Alexander, 1968). Reproduced by permission of Sidgwick and Jackson Ltd.

Figure 3.4 shows the apparatus used by Alexander to study the pressure–volume relationships of the swim bladder. An anaesthetized fish is placed in a flask which is completely filled with weak anaesthetic solution and the flask has a side-arm capillary tube which is partly filled with solution. The pressure on the fish is altered by means of compressed air or a vacuum pump and the pressure is measured by means of a mer-

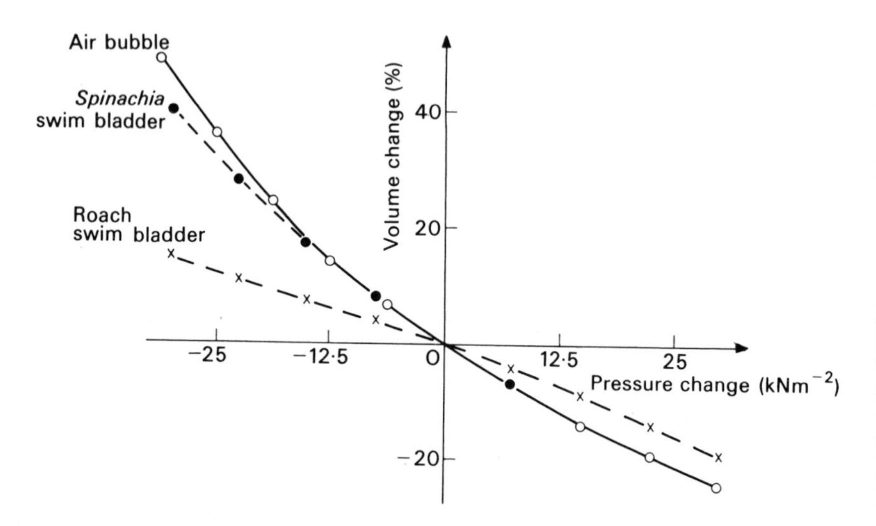

Fig. 3.5. P-V relationships of swim bladder. Changes of volume of an air bubble and of the swim bladders of *Spinachia* and roach (*Rutilus*), caused by changes of external pressure. (From Alexander, 1968). Reproduced by permission of Sidgwick and Jackson Ltd.

cury manometer. The volume change induced by a change in pressure is calculated from the distance which the water meniscus travels along the capillary tube of known cross-sectional area. The results (Fig. 3.5) show that the pressure–volume relationships of the swim bladder of fish of the genus *Spinachia* follow closely the behaviour expected from the gas laws. Fish of this order have in fact swim bladders with slack, highly extensible walls. The *Cypriniformes* (of which the roach is an example) deviate from the air-bubble behaviour. This is because they have swim bladders with taut inextensible walls. This means in fact that when a roach changes its depth its density changes much less than if it had a more extensible swim bladder.

3.5 Work in Changing the Volume of a Fluid

In mechanical systems (Chapter 1, p. 8) when a force **F** acts on the system, the work done is the product of the system's displacement dx and the component of the force F_x along the displacement. The sign convention of mechanics provides for positive work when $F_x\,dx$ is positive, i.e. work done *on* the system is positive. Unfortunately, however, the thermodynamicists of old adopted the opposite sign convention because they focused their attention on the work output of the heat engines they were studying. They said that work done *by* the system is positive and although this is a convention used in many physics texts (including the admirable Sears & Zemansky) it will not be used here because it is confusing. Therefore, our sign convention is *work done on the system is positive.*

Consider a fluid contained in a cylinder equipped with a movable piston (Fig. 3.6a). Suppose the cylinder has cross-sectional area A and that the pressure exerted on the system at the piston face is P. The force exerted on the system is therefore PA. If the piston moves an infinitesimal distance dx, then the work dw done by this force on the system is

$$dw = PA\,dx \qquad 3.9$$

The work done is positive as the displacement dx is in the same direction as the force but

$$A dx = -dV$$

where dV is the change in volume.

The negative sign occurs here as the volume decreases when dx is positive. Hence

$$dw = -PdV \qquad 3.10$$

and in a finite change of volume from V_1 to V_2 the total work done will be given by

$$W = -\int_{V_1}^{V_2} P\,dV \qquad 3.11$$

The total work done on the system is therefore the negative of the area under the pressure–volume curve (Fig. 3.6b).

If the pressure remains constant then

$$W = -P(V_2 - V_1) \tag{3.12a}$$

and positive work is done on the system when $V_1 > V_2$, i.e. when the fluid is compressed.

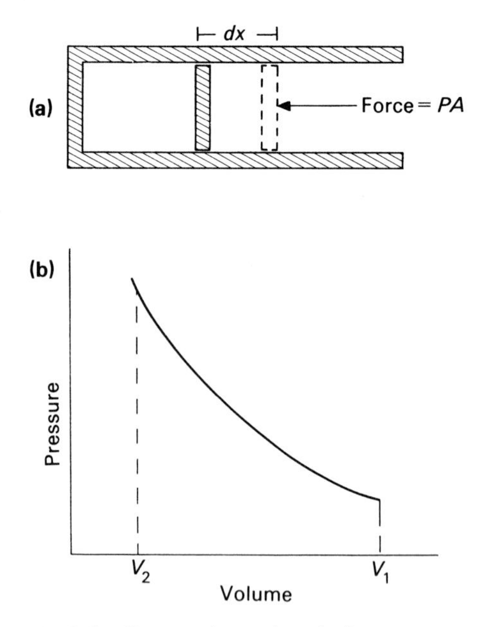

Fig. 3.6. (See text for explanation)

It can also be shown that when the pressure varies, but the volume of fluid remains constant, the work done on the fluid is

$$W = (P_2 - P_1)V \tag{3.12b}$$

3.6 Work Done in an Adiabatic Process

Applying the First Law of Thermodynamics to an adiabatic process ($\Delta Q = 0$) we get from equation 2.2

$$\Delta E = \Delta W \tag{3.13}$$

Thus the change in internal energy of a system in an adiabatic process is equal to the work done. If the work done is positive, e.g. when a system is compressed, then ΔW is positive, ΔE is positive and the temperature of the system increases. If ΔW is negative, e.g. when a system expands, the internal energy decreases and the temperature falls.

An adiabatic expansion is one major cause of precipitation (rainfall). When a parcel of air moves rapidly upwards, there is rapid expansion because of the decrease in pressure. In the rapid process there is no time for the parcel to exchange heat with the environment and so the temperature is lowered. If the temperature is lowered below the dew point, water condenses and rain falls.

The horizontal mixing of two unsaturated air masses can also produce precipitation (Fig. 3.7). Air masses to the right of the curve are unsaturated, whereas those to the left are saturated. *A* and *B* are both unsaturated, but they can combine to form a third mass *C* that is super-saturated at the new temperature and so will become cloudy. If there are condensation nuclei present, rain will fall.

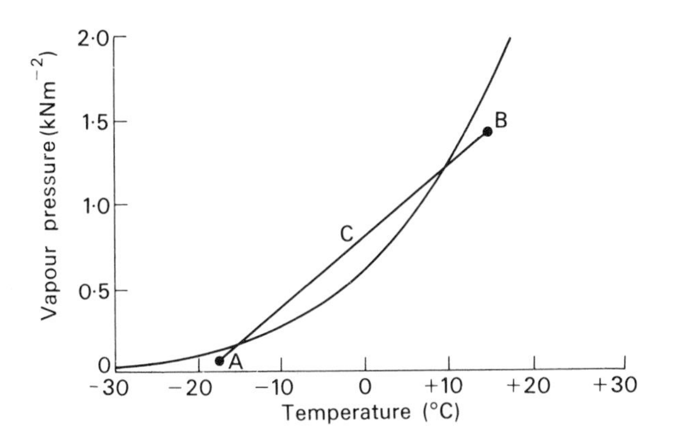

Fig. 3.7. Horizontal mixing of two air masses. The saturation vapour pressure curve as given and the horizontal mixing of two unsaturated air masses *A* and *B* to the right of the curve results in one supersaturated air mass *C*. (After Barry & Chorley, 1968)

Contact cooling is another method of producing a saturated air mass. On a clear winter's night strong radiation from the land will cool the surface of the earth very quickly and this surface cooling will gradually extend to the moist lower air, reducing the temperature to a point where condensation occurs in the form of dew, fog, or frost, depending on the amount of moisture involved and the dew point value. When the latter is below 0°C it is referred to as the hoar frost point and the air is saturated with respect to ice.

3.7 Solubility of Gases in Liquids

The maximal amount of gas that can dissolve in a given liquid depends on three factors.

(i) The pressure of the gas; solubility increases with pressure.

(ii) The temperature of the solvent; the solubility decreases with increasing temperature.

(iii) The solute content of the solvent; the solubility decreases with increasing solute content.

The *absorption coefficient* of a gas in a particular solvent is that volume of gas at STP that can dissolve in unit volume of the liquid at 0°C. The absorption coefficient of CO_2 in water is 1.713 atmospheres^{-1} and this means that 1 litre of water at 0°C dissolves 1.713 litres of CO_2 at STP. The absorption coefficients of oxygen and nitrogen at 0°C are 0.049 atm^{-1} and 0.024 atm^{-1} respectively.

Absorption coefficients can also be given for solvents at temperatures other than 0°C but it should be carefully noted that in these cases the amount of dissolved gas is given by the volume this dissolved gas would occupy at STP (Table 3.2).

Table 3.2

Temperature (°C)	Absorption Coefficient (oxygen in water)
0	0.0489 atm^{-1}
10	0.0380
20	0.0310
30	0.0261

If a gas is present at a pressure of a fraction of an atmosphere, then its solubility will be reduced by the same fraction.

For example, air at sea level contains about 21% oxygen and although this air has a pressure of 10^5 N m^{-2}, the amount of oxygen that dissolves corresponds to a pressure of 2.1×10^4 Nm^{-2}.

We can calculate the solubility of any gas if we know the absorption coefficient a, the partial pressure p, and the pressure P of the total gas mixture.

When $P = 1$ atmosphere (10^5 Nm^{-2})

$$V = \frac{ap}{10^5} \text{ litres per litre} \qquad 3.14$$

and V in this case is the volume in litres of the gas reduced to conditions of STP, that dissolves in 1 litre of solvent

Problem 3.2

Lake Titicaca lies at 14,000 ft between Peru and Bolivia and the barometric pressure is about 59.2 kNm^{-2}. How much oxygen can dissolve in *1 litre* at 20°C? (After Jarman, 1970.)

Answer

(*a*)

$$pO_2 = \frac{21}{100} \times (59.2 - 2.3) = 12 \text{ kNm}^{-2}$$

$$\therefore V = \frac{0.031 \times 12}{10^5}$$

$$= 3.6 \times 10^{-3} \text{ litre}$$

(*b*) Similarly, show that there is about 6.4×10^{-3} litres of oxygen per litre of Norfolk Broad water (at sea level), making the somewhat doubtful assumption that the water contains no salt or dissolved organic material.
Note: atmospheric pressures must be corrected for saturated water vapour pressure.

3.8 Osmotic Pressure

The concept of *osmotic pressure* is a very widely used one, and in order to understand it we have to go to the somewhat artificial system shown in Fig. 3.8.

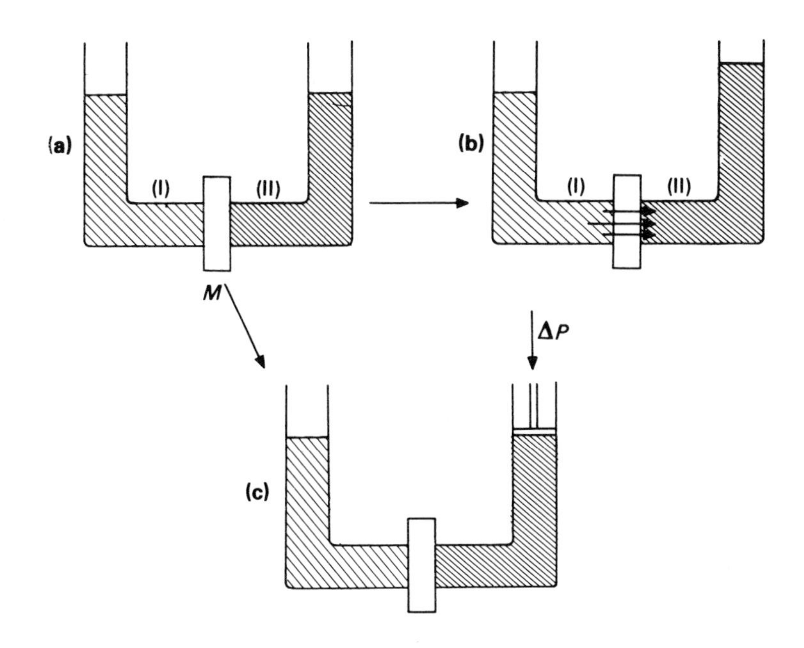

Fig. 3.8. (a) An ideally semipermeable membrane M separates two solutions of different solute concentrations ($C_{II} > C_I$). (b) Experimentally we find that water moves from phase I to II. (c) This net movement can be prevented by applying a pressure on phase II.

A membrane M separates two sucrose solutions, the solution on the right-hand side being more concentrated than that on the left. Initially, the levels in the two arms are at the same height, but if the membrane is what is called a *semipermeable* one, and it allows the passage of water only, and not solute, then the level in the right hand tube is found to rise. This means that there is a flow of water from left to right, and the driving force of this flow is the difference in the *water potential* between the two sides. The water potential of a phase depends on two factors: (i) the concentration of solute dissolved in the phase, and (ii) the hydrostatic pressure on the phase. The presence of dissolved species lowers the water potential, whereas application of a pressure raises the potential.

$$J_W = L_p\,(P_{\mathrm{I}} - P_{\mathrm{II}}) - k(C_{\mathrm{I}} - C_{\mathrm{II}}) \qquad 3.15$$

$$= L_p\,\Delta P - k\,\Delta C \qquad 3.16$$

where J_W is the flow of water; P_{I} and P_{II} are the pressures on the semipermeable membrane in phases I and II respectively; C_{I} and C_{II} are the concentrations of solute in phases I and II respectively; and k and L_p are two constants. The negative sign arises because water flows from a phase where the concentration of dissolved solute is low, to one where it is high.

Now, it can be shown, both from theory and from experiment, that for an ideally semipermeable membrane and ideal solutions

$$k = RT\,L_p$$

Hence equation 3.16 can be written in the form

$$J_W = L_p(\Delta P - RT\Delta C) \qquad 3.17$$

$$= L_p\,(\Delta P - \Delta\pi) \qquad 3.18$$

where $\Delta\pi = RT\,\Delta C$ is called the osmotic driving force or osmotic pressure and it is determined by the sum of *all* the solute molecules and individual ions in the phase; $(P_{\mathrm{I}} - \pi_{\mathrm{I}})$ and $(P_{\mathrm{II}} - \pi_{\mathrm{II}})$ are the water potentials (ψ) of phases I and II respectively. L_p is often referred to as the osmotic permeability.

When the pressure difference between the two phases is adjusted so that there is no net water flow i.e. $J_W = 0$, then

$$\Delta P = \Delta\pi \qquad 3.19$$

and the pressure required to balance the osmotic flow of water is called the *osmotic pressure.*

The *osmolarity* of a solution is the concentration of dissolved, dissociated solute, e.g. a 155 mM solution of NaCl is said to contain 310 milliosmoles per litre or strictly speaking 310×10^6 milliosmoles per cubic metre.

The term *tonicity* is widely used to compare the osmolarity of one solution with another. One solution is said to be hyperosmotic, isosmotic or hyposmotic with respect to another, when it contains a higher, the same, or lower concentration of dissolved solute respectively. A 155 mM solution of NaCl is approximately isosmotic with respect to blood. Raisins swell when soaked in tap water because they are hyperosmotic with respect to the water.

3.9 Osmotic Phenomena in Plants

While most animal cells are isosmotic with the blood that bathes them, many plant cells are in a dilute aqueous environment, pond water for example, where the osmotic pressure is much lower than in the plant cell contents (vacuole and cytoplasm). Osmotic equilibrium can be achieved between the two phases, however, as the thick cellulose plant cell wall exerts a *turgor pressure* (P_T) (Fig. 3.9).

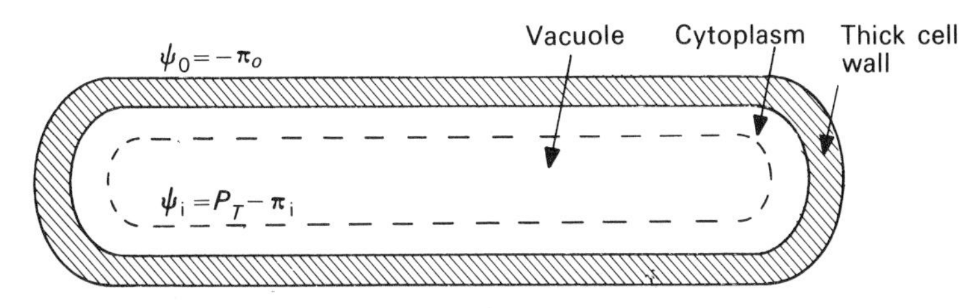

Fig. 3.9. The thick cellulose wall enables the relatively concentrated cytoplasm to remain in osmotic equilibrium with the dilute aqueous environment. (After Thain, 1967).

At equilibrium

$$\psi_i = \psi_o \qquad 3.20$$

i.e.

$$P_T - \pi_i = -\pi_o \qquad 3.21$$

The water potentials of plant cells, or indeed any cell, can be measured by immersing them in a series of solutions of a non-penetrating solute such as sucrose. From equation 3.19 the water potentials of the cells are equal to the osmotic pressures of the solutions in which the cells neither gain nor lose water. Uptake or loss of water may be detected by changes in weights or volumes of the cells.

See also the possible part played by osmosis in phloem translocation (Chapter 4, p. 55).

References

Alexander R.M. (1968) *Animal Mechanics.* Sedgwick and Jackson, London.
Alexander R.M. (1971) *Size and Shape.* Arnold, London.
Barry R.G. & Chorley R.J. (1968) *Atmosphere, Weather and Climate.* Methuen, London.
Burns D.M. & MacDonald S.G.G. (1970) *Physics for Biology and Pre-Medical Students.* Addison-Wesley, London.
Dick D.A.T. (1966) *Cell Water.* Butterworth, London.
Jarman M. (1970) *Examples in Quantitative Zoology.* Arnold, London.
Sears F.W. & Zemansky M.W. (1964) *University Physics.* Addison-Wesley, Reading, Mass.
Thain J.F. (1967) Principles of Osmotic Phenomena. *Royal Institute of Chemistry, Monographs for Teachers No. 13.*

Chapter 4
Fluid Flow and Viscosity

4.1 Introduction

Fluids flow as the result of driving forces, e.g., in the case of a pipe, flow occurs when there is a pressure difference across the ends of the pipe. This simple basis is the starting point for investigation into systems as widely diverging as blood flow through veins and arteries and phloem transport in plants. In this discussion we shall be concerned initially with *ideal fluids* which are considered to be *incompressible*.

4.2 The Continuity Equation

Consider a fluid flowing along a tube whose cross-sectional area varies from point to point (Fig. 4.1). At position (1) the cross-sectional area is A_1 and the linear velocity is v_1, while at (2) they are A_2 and v_2 respectively. As the fluid is considered to be incompressible, the density ρ will not vary.

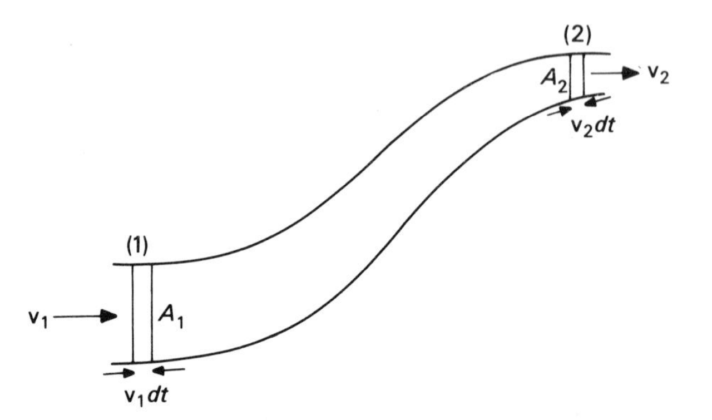

Fig. 4.1. Flow through a tube of varying cross-sectional area. At positions (1) and (2) the cross-section areas are A_1 and A_2 and the velocities v_1 and v_2 respectively.

Now, as no fluid leaves through the walls, the mass flowing in across A_1 in a time dt must equal that leaving A_2 in the same time. Hence

$$\rho A_1 v_1 \, dt = \rho A_2 v_2 \, dt$$

and so

$$A_1 v_1 = A_2 v_2 \qquad 4.1$$

This obviously implies that as the vessel narrows, the linear velocity of the fluid through it increases.

Problem 4.1

Does the continuity equation apply to the following data obtained for the cardiovascular system in man?

	Total Cross-sectional Area of Vessels (m^2)	Linear Velocity of Blood ($m s^{-1}$)
Aorta	2.5×10^{-4}	30×10^{-2}
Capillary bed	1.9×10^{-1}	5×10^{-4}
Vena Cava	1×10^{-3}	8×10^{-2}

4.3 Bernoulli's Equation

When an incompressible fluid flows along a tube of varying cross-section its velocity changes. It must therefore be acted on by a resultant force, and this means in fact that pressure may vary along the tube. Let us consider the general case where the height of the tube above some reference level also changes (Fig. 4.2).

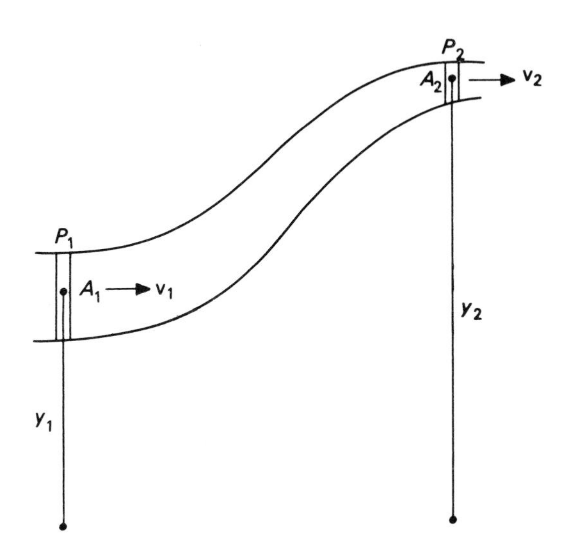

Fig. 4.2. The flow through a tube of varying cross-sectional area as in Fig. 4.1, and y_1 and y_2 are the heights of the tube above some arbitrary reference level.

Consider an element of the fluid at positions (1) and (2). The pressures at the two positions are P_1 and P_2 respectively, y_1 and y_2 are the heights of the centres of fluid mass above the earth's surface, and v_1 and v_2 are the velocities of the fluid. The net work done on moving an element of volume V from position (1) to (2) is $(P_1 - P_2)\,V$ (equation 3.12b) and this must equal the change in potential and kinetic energies of the element, i.e.

$$(P_1 - P_2)\,V = mgy_2 - mgy_1 + \tfrac{1}{2}\,mv_2{}^2 - \tfrac{1}{2}\,mv_1{}^2$$

as $$V = m/\rho$$

then $$P_1 - P_2 = \rho g y_2 - \rho g y + \rho v_2{}^2 - \rho v_1{}^2$$

or $$P + \rho g y + \tfrac{1}{2}\rho v^2 = \text{constant} \qquad 4.2$$

In most problems of biological interest, $y_1 = y_2$ and Bernoulli's equation becomes

$$P + \tfrac{1}{2}\rho v^2 = \text{constant} \qquad 4.3$$

As the velocity increases with decreasing area (equation 4.1), equation 4.3 implies that the pressure in a tube decreases at a point of constriction (Fig. 4.3) and this has important consequences in blood vessels.

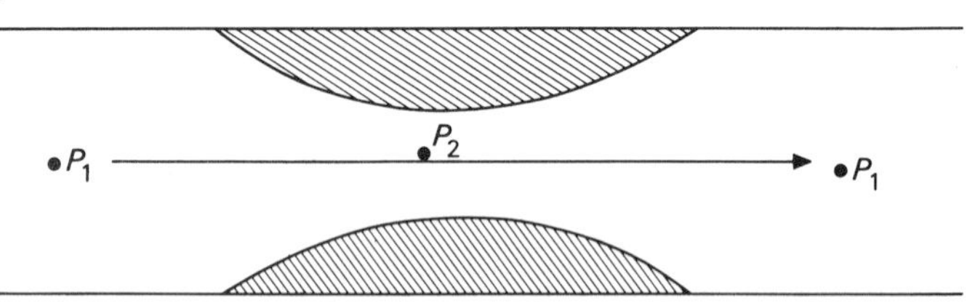

Fig. 4.3. The pressure at a point within the constricted area is less than the pressure in the normal vessel ($P_2 < P_1$) and this causes flutter, a medical condition.

In equation 4.2, $\tfrac{1}{2}\rho v^2$ has the units kinetic energy per unit volume. This expression is sometimes referred to as the *kinetic energy density* of the fluid and it also has the units of pressure.

So far it has been possible to derive two very valuable equations that apply to the gross movements of fluids. However, we cannot for example analyse in detail the flow through capillaries without taking into account the fluid viscosity.

Problem 4.2

(*a*) The mean volume flow of blood in the arterial system is $8.5 \times 10^{-5}\ \mathrm{m^3\,s^{-1}}$. If the area of the aorta is $3 \times 10^{-4}\ \mathrm{m^2}$, find the linear velocity of flow and show that the average kinetic energy per unit volume is approximately $45\ \mathrm{J\,m^{-3}}$. Show that the pressure corresponding to this kinetic energy per unit volume is only a small fraction of the overall arterial pressure (ref. Problem 4.1).

(*b*) At the beginning of the heart's ejection period, the linear velocity of the blood may be 3 times the average velocity. Show that this explains the experimentally measured pressure difference of some 200 $N m^{-2}$ between the *aorta*, into which the blood is rushing, and the *left ventricle*, where the velocity is near zero, during this short period.
(*c*) Show that during heavy exercise, when the cardiac output is 5 times normal, the kinetic energy created becomes a significant fraction of the total work done by the heart.

Problem 4.3
In diseased arteries, an atherosclerotic plaque can narrow the lumen to under one fifth of the original diameter (Fig. 4.3). By how much does this reduce the arterial pressure within the constricted area?

If in fact the arterial pressure is lowered below the critical closing pressure, the artery will close (Fig. 4.4). However, when the flow is reduced to zero, the kinetic

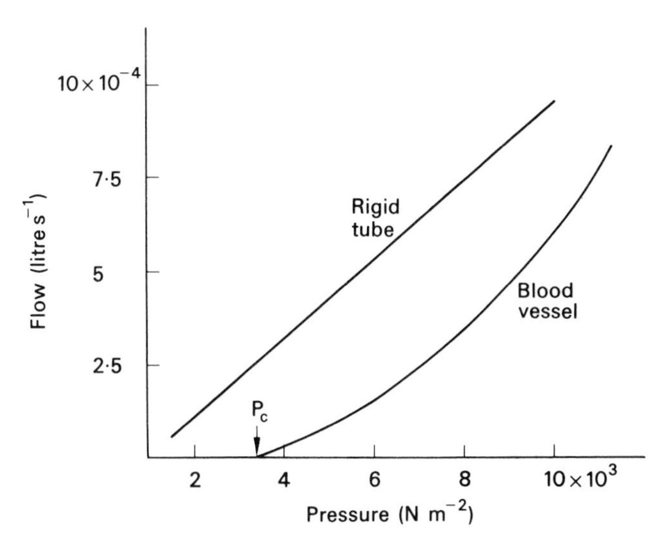

Fig. 4.4. The relationship between pressure and flow in an artery. It is not linear because of the elastic properties of the walls, and a certain pressure P_c is required to overcome the elasticity and open the vessel.

energy component disappears, the pressure builds up once more and the cycle is repeated. This sequence of events is called *flutter*, and the sounds produced by this vibratory motion are used diagnostically.

4.4 Pressure and Flow in Tubes

In order to drive a constant flow of liquid between two positions along a river or along a pipe for example a constant pressure difference must be maintained in order to overcome the internal friction or *viscosity* of the fluid. Newton was the first to ascribe the viscosity to a lack of slipperiness as the different parts of the fluid passed one another. This hypothesis lead to the concept that the movement of fluid is effected by infinitesimally thin laminae sliding over one another.

The regularity of fluid movement can be seen if a series of straws are dropped in a line across a smooth flowing river, and their positions are noted at successive times (Fig. 4.5a). It is found that the straws in the middle move furthest while those nearest the bank scarcely move at all. The straws map out the velocity profile of the flow at the surface and it consists of *streamlines*, i.e. lines joining points of constant velocity, running parallel to the banks. Adjacent streamlines have different velocities and so a force has to be exerted to allow the lines to slip over one another. Regions of constant

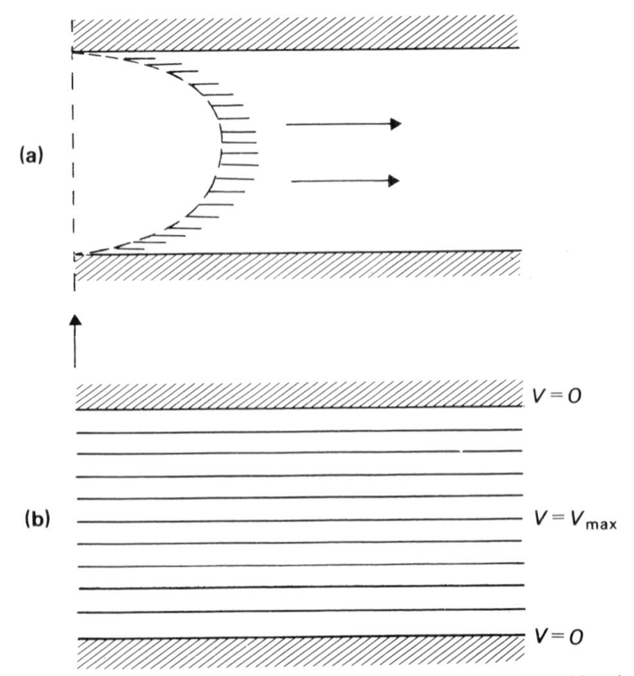

Fig. 4.5.(a) Straws dropped into a river along a line indicated by the vertical arrow map out the profile of the velocity of flow at the surface. The profile is *parabolic* in form. (b) Hypothetical streamlines of a flow at the surface of a river. The streamlines at the bank have zero velocity while the streamline in the centre has a maximal velocity.

velocity extend into the bulk of the fluid making up laminae that carry the flow of liquid. In a river, the shape of the laminae depend on the profile of the river bed, but in a cylindrical tube, the laminae would simply resemble telescope tubes. The laminae near the solid surface would scarcely move, while those at the centre would be moving with maximal velocity.

One way of determining the viscosity of a fluid is to measure the force required to maintain a certain velocity between two plates in a fluid. It is found experimentally that the force required depends on the area of the plates, the *velocity gradient* between them and of course the fluid viscosity (Fig. 4.6). The relationship is expressed in the equation

$$\mathbf{F} = \frac{A\mathbf{v}\zeta}{d} \qquad 4.4$$

where $\mathbf{v}/d$ is the velocity gradient between the two plates and ζ, the viscosity, has dimensions $ml^{-1}\ t^{-1}$.

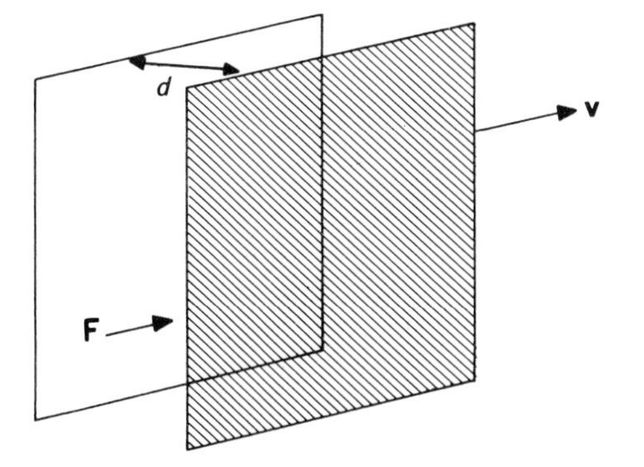

Fig. 4.6. The force **F** required to move one plate past the other in a viscous medium depends on the velocity gradient $\mathbf{v}/d$, the area of the plates A and the viscosity of the medium.

Equation 4.4 applies equally well to fluid laminae and as they are considered to be infinitesimally thin, the equation takes the differential form

$$\mathbf{F} = A\zeta\frac{d\mathbf{v}}{dx} \qquad 4.4a$$

To find the form of the velocity profile in a system we have to integrate this equation. We shall consider the case of most use in biology, namely that of a cylindrical tube, e.g. a blood vessel.

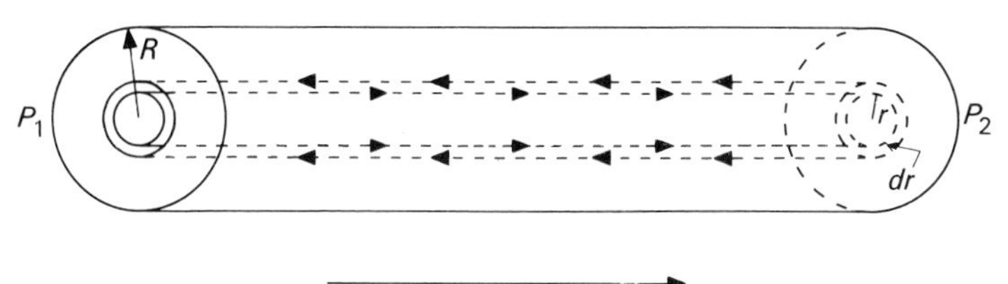

Fig. 4.7. A tube of radius R has a pressure difference $P_1 - P_2$ across its ends. The diagram shows the viscous forces on a lamina moving through the tube.

Footnote
The mathematical treatment given here is slightly more complex than that normally found in elementary texts, but I believe that it is more rigorous as it treats the case of a lamina moving in equilibrium under the action of the pressure forces rather than that of a solid cylinder of fluid (cf. Sears & Zemansky, 1964).

Consider a tube of radius R (Fig. 4.7) containing a fluid flowing from left to right under the action of a pressure difference $P_1 - P_2$ ($P_1 > P_2$) across its ends a distance l apart. Fluid is carried along the tube by an infinitesimal number of laminae sliding past one another. The velocity is constant within each of these laminar tubes. The velocity is considered zero at the tube walls and further it is assumed to have the general profile indicated in Fig. 4.5a.

Consider further the tubular lamina of inner and outer radii r and $r + dr$ respectively. The velocity within the lamina is constant and the velocities of the fluid adjacent to the inner and outer surfaces are $\mathbf{v}_r$ and $\mathbf{v}_{r+dr}$ respectively (and $\mathbf{v}_r > \mathbf{v}_{r+dr}$). The force driving the fluid to the right along a length l of tube is the product of the pressure difference $(P_1 - P_2)$ and the cross-sectional area of the lamina $(2\pi r\, dr)$. The viscous force at the outer surface of tube (F_{r+dr}) is to the left as the velocity of the lamina is greater than the velocity of the surrounding fluid. F_{r+dr} is equal to $-\zeta 2\pi(r + dr)\, l\, (d\mathbf{v}_{r+dr}/dr)$ (the negative sign occurs because $\mathbf{v}$ decreases as r increases). Similarly the frictional force (F_r) at the inner surface is to the right and is equal to $+ \zeta 2\pi r\, l(d\mathbf{v}_r/dr)$.

At equilibrium, the pressure driving force to the right equals the net frictional force to the left, i.e.

$$(P_1 - P_2)\, 2\pi r\, dr = -\zeta\, 2\pi(r + dr)\frac{l\, d\mathbf{v}_{r+dr}}{dr} + \zeta\, 2\pi r\frac{l\, d\mathbf{v}_r}{dr} \qquad 4.5$$

but $$\frac{d\mathbf{v}_{r+dr}}{dr} = \frac{d\mathbf{v}_r}{dr} + \frac{d^2\mathbf{v}_r}{dr^2}\, dr_r \qquad \text{(Ferrar, 1967, Chapter 7)} \qquad 4.6$$

$$\therefore \quad (P_1 - P_2) r\, dr = -\zeta\, l\frac{d\mathbf{v}_r}{dr}\, dr - \zeta l r\frac{d^2\mathbf{v}_r}{dr^2}\, dr \qquad 4.7$$

(ignoring second order terms in dr)

$$= -\zeta l\frac{d}{dr}\left(r\frac{d\mathbf{v}}{dr}\right)dr$$

and on integration

$$\frac{(P_1 - P_2)}{2}\, r^2 = -\zeta l r\frac{d\mathbf{v}}{dr} + \text{constant} \qquad 4.8$$

Because of the general form of the profile (Fig. 4.5a) $d\mathbf{v}/dr$ is zero when r is zero and hence the integration constant is zero. To obtain the exact form of the velocity profile, equation 4.8 must be integrated. The limits of integration chosen are from the laminar tube to the pipe wall because we know that the velocity at the wall is zero.

$$\int_{\mathbf{v}_r}^{0} d\mathbf{v} = -\frac{(P_1 - P_2)}{2\zeta l}\int_r^R r\, dr \qquad 4.9$$

and so $$\mathbf{v}_r = \frac{P_1 - P_2}{4\zeta l}(R^2 - r^2) \qquad 4.10$$

This is in fact the equation of a parabola. To a biologist a picture of the velocity profile of fluid moving through a pipe is not of immediate interest, but it can be used to obtain a value for the volume flow through a vessel.

To find the volume flow across a cross-section of the tube we have to perform another integration as the velocity varies radially. We consider the volume flow across a small cross-sectional area (Fig. 4.8) a distance r from the centre where the velocity is constant and equal to v_r, i.e. across the shaded area shown above.

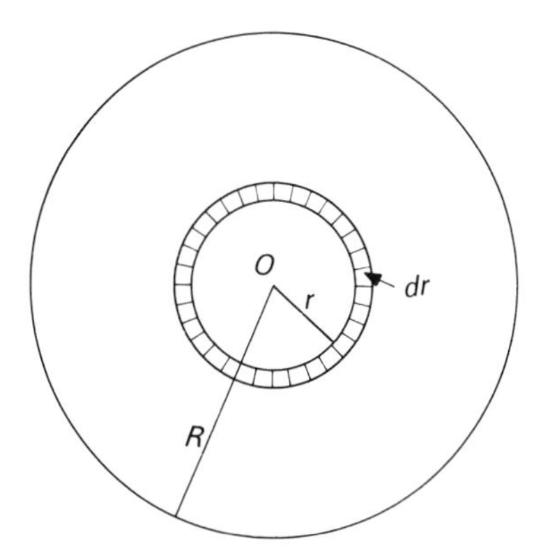

Fig. 4.8. The volume of fluid passing across the shaded area in a time dt is given by $2\pi r\, dr\, v\, dt$, where v is the linear velocity of flow.

The volume of fluid (dQ) crossing this small area dA in time dt will be given by $v dA$ which is equal to $v 2\pi r\, dr\, dt$. If we integrate across the whole cross-sectional area, i.e. from O to R then we can find the rate of flow in $m^3\ s^{-1}$ through the total cross-section

$$\frac{dQ}{dt} = \int_0^R v_r 2\pi r\, dr \qquad 4.11$$

and v_r given by equation 4.10

$$\text{Hence} \quad \frac{dQ}{dt} = \frac{\pi(P_1 - P_2)}{2\xi l} \int_0^R (R^2 - r^2)\, r\, dr \qquad 4.12$$

$$\frac{dQ}{dt} = \frac{\pi}{8} \frac{R^4}{\xi} \frac{P_1 - P_2}{l} \qquad 4.13$$

$$\text{but} \quad \frac{dQ}{dt} = Av \qquad 4.14$$

where A is the cross-sectional area of the tube and v, the linear velocity of flow through it.

$$\text{Hence} \quad v = \frac{R^2}{8\xi} \frac{P_1 - P_2}{l} \qquad 4.15$$

Equation 4.13 was first derived empirically by a physician Poiseuille from his observations on blood flow in the capillary bed of animals. $(P_1 - P_2)/l$ is called the pressure gradient along the tube. Although the equation predicts a linear relationship between blood flow and pressure gradient, this is not found as the blood vessels are composed of the elastic elements, elastin and collagen, which stretch under pressure and so the vessel offers less resistance to flow than if it were rigid. Because the vessels are in a continual state of stress due to the elastic forces, there is a certain minimum pressure required to keep the tube open (Fig. 4.3).

There is also a fundamental objection to the application of the Poiseuille equation to fluids that have very large particles suspended in them. The viscosity of a fluid involves the concept of a velocity gradient measured across infinitesimally small laminae and although such a concept is accurate for a homogeneous fluid where all the particles are of molecular size, it cannot be true of blood, for example, where the laminae can take on no smaller dimensions than the thickness of the red cells. The viscosity of blood in fact depends on the concentration of the blood cells and when this is abnormally high (in polycythemia), the viscosity of the blood can increase to as much as five times that of normal blood, imposing a similar increase in the resistance to flow of blood through the arterial system.

4.5 Reynold's Number

When the velocity of a fluid flowing in a tube exceeds a certain critical value the flow is no longer laminar, but turbulent (Fig. 4.9).

Experiments have shown that a combination of four factors determine whether

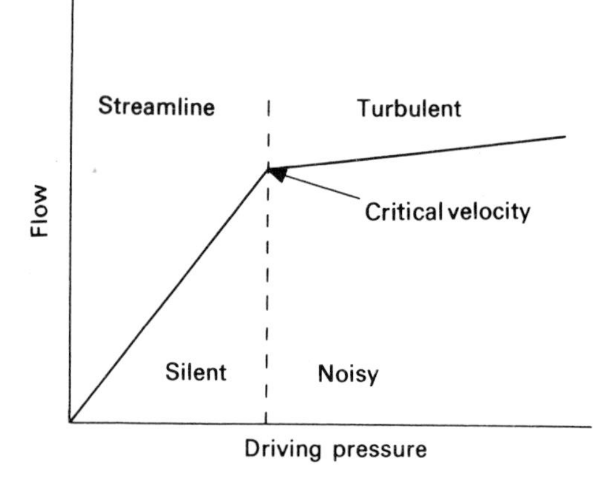

Fig. 4.9. When Reynold's number is exceeded the flow becomes turbulent and noisy and so some of the driving pressure is lost.

flow is laminar or turbulent. The combination, called Reynold's number N_R, is given by

$$N_R = \frac{\rho v D}{\zeta} \qquad 4.16$$

where ρ is the density of the fluid; v the flow velocity; D is the diameter of the tube; and ζ the viscosity of the fluid. N_R is a dimensionless quantity, and experiments have shown that when N_R is less than 2000 the flow is laminar and when N_R is greater than 3000 it is turbulent.

Problem 4.4

(*a*) The resting velocity of flow through the aorta of diameter 2×10^{-2} m is 0.3 ms^{-1}. If the density and viscosity of blood are 10^3 kg m^{-3} and 4×10^{-3} kg m^{-1} s^{-1} respectively, show that the flow is laminar.

(*b*) Show that during heavy exercise this need not necessarily be so. (See Problem 4.2.)

4.6 Phloem Transport

Problem 4.5

The rate of sugar transport per unit area of phloem into fruits is of the order of 0.07 mole $metre^{-2}$ sec^{-1}. The average concentration of the sucrose solution moving through the phloem is 300 mM (300 Mm^{-3}).

(*a*) Using the structural data given in Fig. 4.10, calculate the linear velocity of flow through the sieve pores and sieve lumen and hence the pressure required to drive this flow through 1 m length of phloem, i.e. calculate the successive pressure drops due to viscosity along each section of tube lumen and sieve plate.

(*b*) Hence show that the total osmotic driving pressure that could be developed by a difference of concentration of 300 mM sucrose across the phloem membranes, at position A, is inadequate to drive this flow. (Assume that the membranes at A are impermeable to sucrose.)

(*c*) Can you suggest any other driving forces? A good reference to start your research into the controversial phloem transport field is Aikman & Anderson (1971) and for the physiological background see Richardson (1969). The viscosity of the sugar solution is 2×10^{-3} N m^{-2} s.

4.7 Stokes' Law

When a sphere of radius r moves through a stationary fluid and if the motion is non-turbulent, then the viscous drag F on the sphere is given by Stokes' Law

$$F = 6\pi\zeta r v \qquad 4.17$$

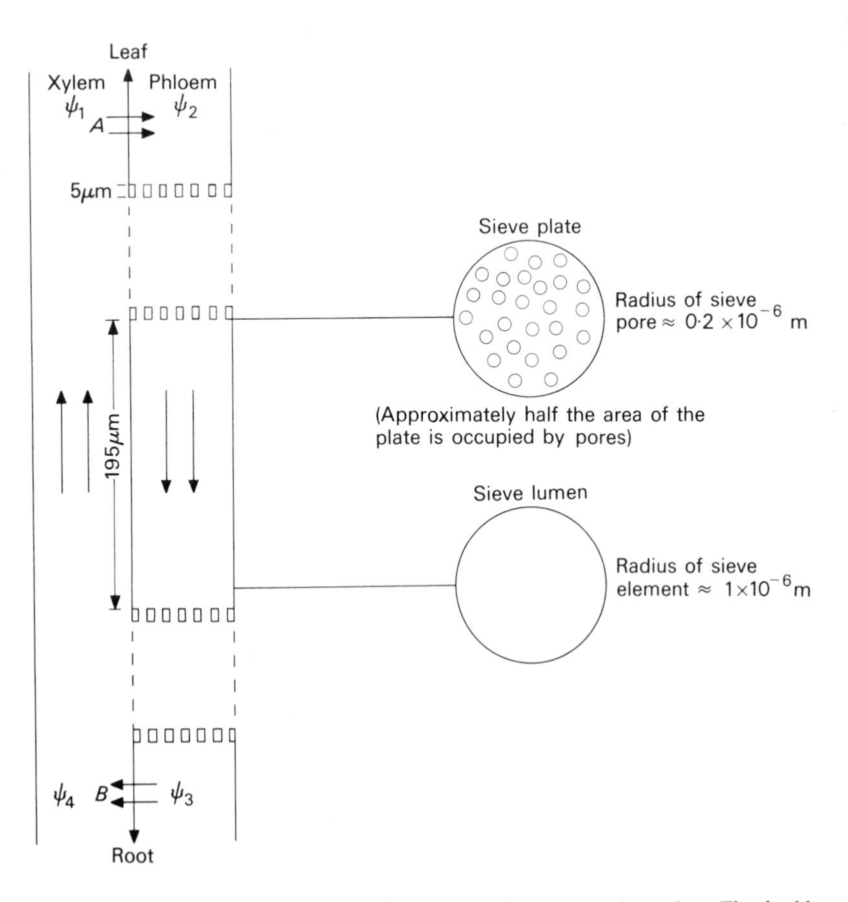

Fig. 4.10. Diagrammatic section of phloem tube and accompanying xylem. The double arrows indicate the direction of bulk flow of sucrose solution. The water potentials are denoted by ψ (Chapter 3). It has been suggested that the main driving force on the water movement is the secretion of sucrose into the upper phloem (position A) and the sucrose concentration excess is believed to be in the region of 300 mM. At A $\psi_1 > \psi_2$ because of the high sucrose concentration in the phloem and at B $\psi_3 > \psi_4$ because of the higher hydrostatic pressure of the phloem. (See also Richardson, 1969)

When such a sphere falls in a viscous medium it reaches a terminal velocity when the retarding forces, viscosity and buoyancy, equal the weight of the sphere. The weight of the sphere is $\frac{4}{3}\pi r^3 \rho_s g$ and the buoyancy force equal to the weight of fluid displaced, is $\frac{4}{3}\pi r^3 \rho_f g$ where ρ_s and ρ_f are the density of the sphere and fluid respectively. When the terminal velocity v_T has been reached

$$\frac{4}{3}\pi r^3 \rho_s g = \frac{4}{3}\pi r^3 \rho_f g + 6\pi \zeta r v_T$$

hence $$v_T = 2r^2 (\rho_s - \rho_f)\, g/9\zeta \qquad 4.18$$

The terminal velocity is also called *sedimentation velocity* when applied to the centrifugation of macromolecules (Fig. 4.11)

$$v_T = \omega^2 R \, 2 r^2 (\rho_s - \rho_f)/9\zeta \qquad 4.19$$

where ω is the angular velocity of the centrifuge tube and R is the distance of the molecule from the axis of rotation. The fact that R appears in the equation shows

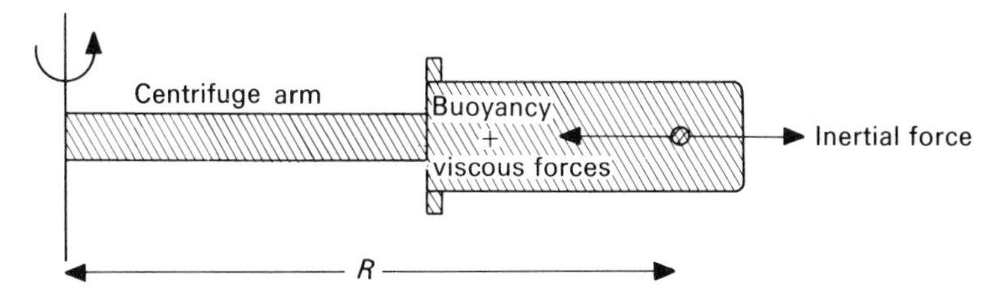

Fig. 4.11. Movement of a molecule in a centrifuge tube. The molecule eventually reaches a terminal, velocity v_T, determined by the resultant of the buoyancy, viscous, and inertial forces.

that it is meaningless to quote r.p.m. alone when quoting centrifuge data; in fact, the number of *g*'s in $\omega^2 R$ is the usual means of indicating the conditions under which macromolecules are sedimented.

References

Aikman D.P. & Anderson W.P. (1971) A Quantitative Investigation of a Peristaltic Model of Phloem Translocation. *Annals of Botany* 35, 761-772.
Alexander R.M. (1968) *Animal Mechanics.* Sedgwick and Jackson, London.
Alexander R.M. (1971) *Size and Shape.* Arnold, London.
Ferrar W.L. (1967) *Calculus for Beginners.* Clarendon, Oxford.
Randall, J. E. *Elements of Biophysics.* Year Book Medical Publishers, Chicago.
Richardson M. (1969) *Translocation in Plants.* Arnold, London.
Ruch T.C. & Patton H.D. (1965) *Physiology and Biophysics.* Saunders, Philadelphia.
Sears F.W. & Zemansky M.W. (1964) *University Physics.* Addison-Wesley, Reading, Mass.
Setlow R.B. & Pollard E.C. *Molecular Biophysics.* Pergamon, London.

Chapter 5
Surface Tension

5.1 Introduction and Definitions

In a homogeneous fluid, the molecules in the body of the fluid have on average no net forces acting on them as their nearest neighbours pull equally in all directions. At the surface of the fluid, however, there is a net force on a molecule which acts towards the bulk phase (Fig. 5.1). The net effect of these inward pulling forces is called the *surface tension.*

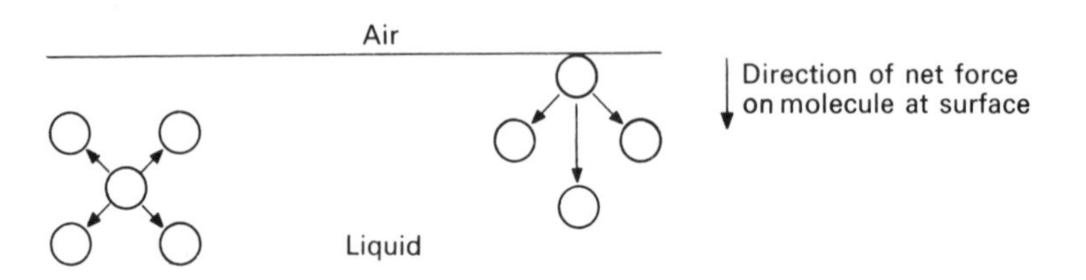

Fig. 5.1. Surface tension forces arise because a molecule at the surface experiences a net force acting towards the bulk fluid.

When a wire ring carrying a loop of thread is dipped in a soap solution and withdrawn, the soap forms a film on the ring and the thread lies slack in the film (Fig. 5.2a). If, however, the film within the thread is punctured the loop is pulled taut by the action of the soap molecules in the bulk phase (Fig. 5.2b).

Fig. 5.2. (a) When the soap film completely covers the loop the thread lies slack in the film. (b) When the film in the centre is broken, however, the thread is pulled taut by surface tension forces.

The forces involved in surface phenomena at a soap-air interface can be measured by means of a slider of mass m_1 and length l which can move easily up and down the arms of a wire bent in the shape of a U (Fig. 5.3). In this device there are two interfaces along which surface tension forces act on the slider, i.e. on the front and back surfaces (Fig. 5.3b). When the wire is dipped in solution with the slider at the bottom, it is rapidly pulled to the top under the action of the surface tension forces.

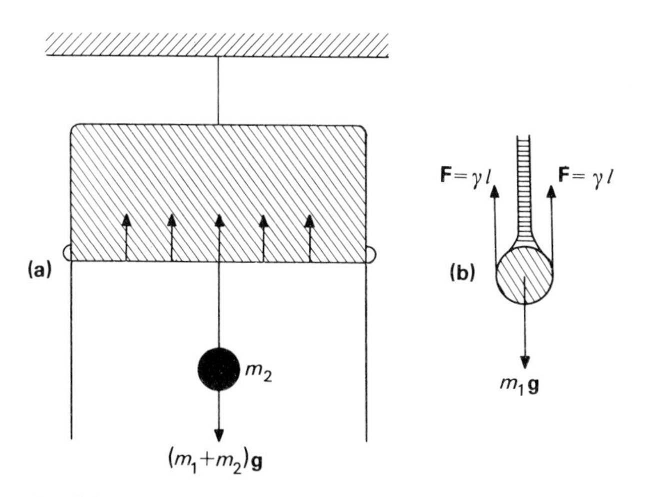

Fig. 5.3. (a) Surface tension forces at the soap-air interface pull the slider up and the forces can be measured by hanging weights on the slider. (b) Cross-section through wire slider and soap film. The film has two interfaces, each of length l and hence the total surface tension force upwards equals $2\gamma l$.

The slider can be held in equilibrium by attaching small weights m_2, and it is found that this weight will hold the slider at any position on the wire. This means that the surface forces, which are equal to $(m_1 + m_2)\,g$ are independent of the total area of the film. Hence surface phenomena differ from elastic ones as in the latter case, stretching a body requires greater forces to maintain equilibrium (Chapter 1, p. 17). In surface phenomena the molecules are not stretched, but the surface in fact expands at the expense of molecules from the bulk phase.

The surface tension forces depend only on the molecular species on either side of the interface, and on the length of the interfaces. The total length of the soap-air interface is $2l$ (Fig. 5.2b) and the surface tension γ is defined as the force acting perpendicular to the interface, divided by the total length of the interface, i.e.

$$\gamma = \frac{(m_1 + m_2)\,g}{2l} = \frac{F}{2l} \tag{5.1}$$

and surface tension has the units N m^{-1}. Table 5.1 gives the surface tension of some common liquids when they are in contact with air.

5.2 Surface Energy

Although no extra force is required to expand the area of the soap film, work is done when the area is increased. If the slider is moved a distance y, then the work done is Fy and the total increase in surface area is $2ly$. The work done per unit area in expanding the film is therefore

$$Fy/2ly = F/2l = \gamma \qquad 5.2$$

Hence the surface tension is also the potential energy stored in unit area of surface.

In the same way as mechanical systems tend to move to a position of minimum potential energy and chemical systems to a position of minimum chemical energy, surfaces seek an arrangement of minimal surface energy. In a solution, this can be attained by the formation of a mass of minimum surface area, such as a sphere. Soaps, lipids, and bile salts, for example, tend to lower the surface tension at an air-water interface and so they will tend to accumulate at the interface because their presence lowers the surface energy.

Bile salts aid the digestion of fats by lowering the surface tension at the fat-water interface. This permits an increase in the total area of interface and so one large fat globule can break up into several small ones allowing for a greater area of attack by pancreatic lipases in the small intestine.

Table 5.1

Liquid in contact with air	t°C	Surface tension (Nm^{-1})
Benzene	20	29×10^{-3}
Soap solution	20	about 25×10^{-3}
Bile Salt solution	20	about 20×10^{-3}
Water	20	72×10^{-3}

5.3 Contact Angle

In many situations there are three surfaces where tension forces interact, for example in the rise of a water meniscus in contact with glass (Fig. 5.4). The liquid-air interface lies at an angle θ to the glass wall, and θ is called the *contact angle*. When θ is less than

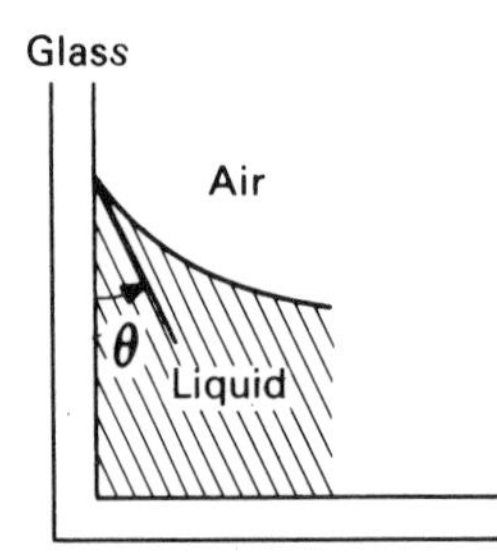

Fig. 5.4. At a water-air interface in a glass vessel the contact angle is less than 90° and so the water is said to wet the glass.

90°, then the liquid is said to *wet* the solid and this is the case illustrated. The contact angle between mercury and glass is greater than 90° and so when a capillary tube is pushed into mercury the surface within the tube is lower than that outside. The contact angle between pure water and paraffin is 110°, so water does not wet paraffin. If detergents are added to the water, the contact angle is reduced and the water will then wet the paraffin (Fig. 5.5).

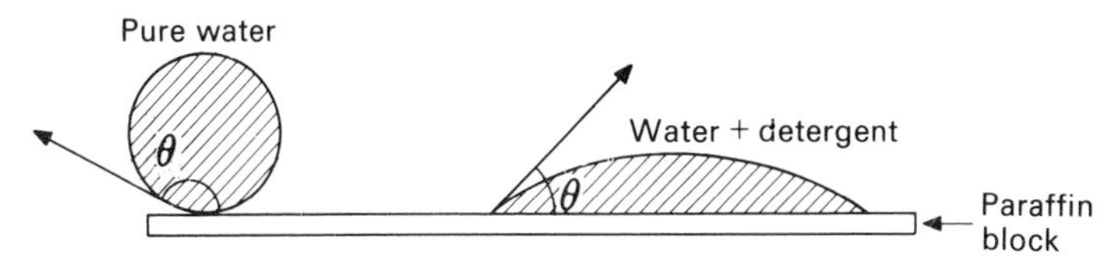

Fig. 5.5. If a detergent is added to the water, it will wet the paraffin.

5.4 The Laplace Equation

Very often we have to calculate the forces acting on spheres, or hemispheres in contact with air. Consider the surface tension forces on a bubble of air in a liquid. The right hand half of the sphere is shown in the diagram (Fig. 5.6a) and it is attracted to the left hand by surface tension forces

$$F = \gamma l = \gamma 2\pi r$$

where r is the radius of the sphere. The force acting in the opposite direction is derived from the pressure difference across the bubble, and the component acting anti-parallel to the surface tension forces is the pressure difference acting on the projected area of πr^2 (Fig. 5.6b).

The total force to the right is $(P_1 - P_2)\pi r^2$. Equating the two forces

$$(P_1 - P_2)\pi r^2 = 2\pi r\gamma \quad 5.3$$

$$P_1 - P_2 = \frac{2}{r}\gamma \quad 5.4$$

or

$$\Delta P = \frac{2}{r}\gamma$$

The collapse pressure on the bubble, i.e. pressure ΔP required to keep the bubble in equilibrium, increases as the radius decreases.

Laplace's equation can be used to calculate the rise of liquids in narrow tubes (Fig. 5.7). At a glass-air interface the contact angle is near zero and so the meniscus can be considered as tangential at the glass and hemispherical in the tube lumen. The collapse pressure serves to drag a column of water upwards as it is not balanced by interactions with an opposing hemisphere.

The upward force will be the collapse pressure times the projected area and this will be balanced by the downward force due to the action of gravity on the fluid mass, i.e.

$$\frac{2\gamma\pi r^2}{r} = mg \qquad 5.5$$

now $m = \rho\, V = \rho\pi r^2 h$; where V is the volume of the fluid; ρ is density; and h the height to which it rises. Hence

$$h = \frac{2\gamma}{\rho r g} \qquad 5.6$$

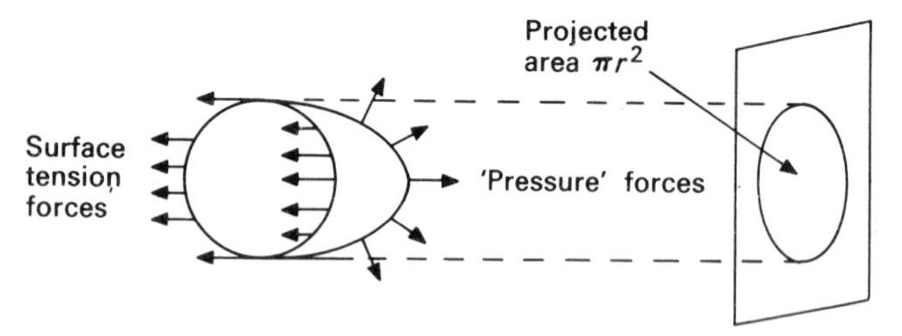

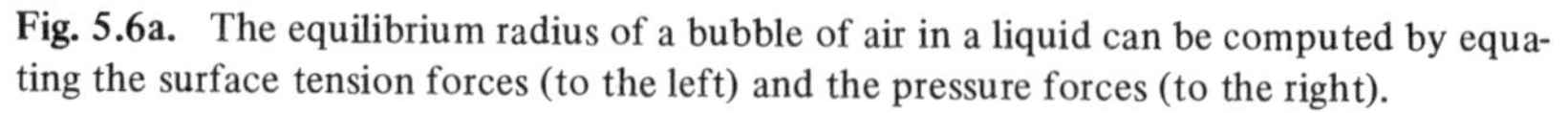

Fig. 5.6a. The equilibrium radius of a bubble of air in a liquid can be computed by equating the surface tension forces (to the left) and the pressure forces (to the right).

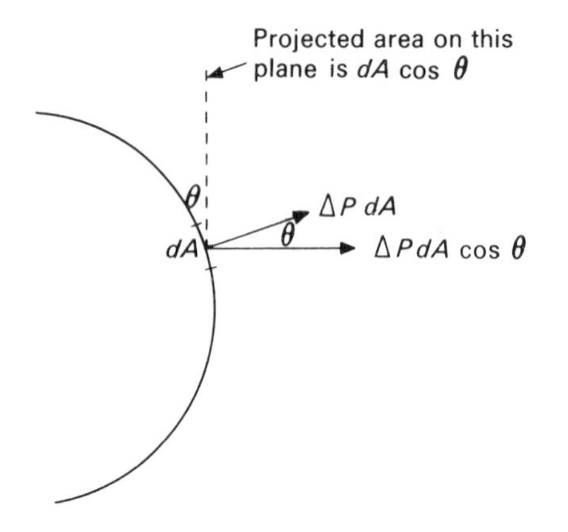

Fig. 5.6b. Consider an element of area dA which makes an angle θ with the vertical plane indicated by the dotted lines. The force normal to this area is $\Delta P dA$, and the component of this force antiparallel to the surface tension forces is $\Delta P dA \cos\theta$. However, $dA \cos\theta$ is the projected area of dA of the vertical plane. Therefore the pressure forces opposing the surface tension forces are given by

$$\sum_{\text{all elements}} \Delta P dA \cos\theta \quad \text{and this equals} \quad \Delta P \sum_{\text{all elements}} dA \cos\theta$$

which in turn equals $\Delta P \times \pi r^2$
where πr^2 is the total projected area of the hemisphere in the plane indicated.

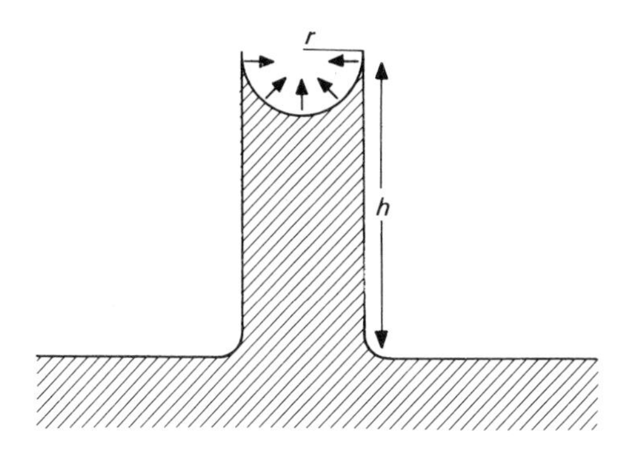

Fig. 5.7. The fluid in a narrow tube is pulled up the by collapse pressure on the hemisphere at the liquid-air interface.

Problem 5.1
Sap flows upwards through the xylem in trees. Diameters of vessels range from 20-400 μm. Can capillarity alone account for the sap rising to a height of 100 metres in some trees, e.g. *Sequoia*? If not, suggest other mechanisms by which this height might be reached. (See Richardson, 1969.) Take density and surface tension of sap as 10^3 kg m^{-3} and 70 x 10^{-3} Nm^{-1} respectively.

5.5 Surface Balance

The instrument used in biophysics laboratories to measure surface tension quantitatively is the surface balance (Fig. 5.8). It consists of a water-filled Teflon trough with a movable Teflon barrier. At one end a platinum strip dips in the water and it is

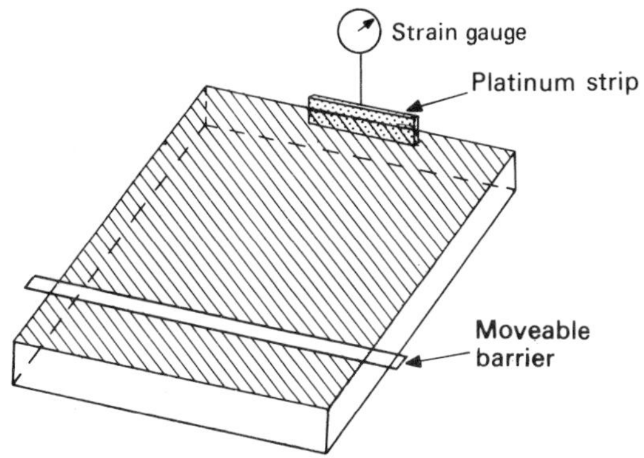

Fig. 5.8. The surface balance. Surface tension at the liquid-air interface pulls on the platinum strip and gives a reading on the calibrated strain gauge.

attached to a calibrated strain gauge which gives a direct measure of the surface tension force pulling down on the strip. Water alone produces a pull of 70 x 10^{-3}Nm^{-1}. When detergent is added the surface tension is reduced to about 30 x 10^{-3} Nm^{-1} and the tension does not change as the barrier moves back and forth. When phospholipids, e.g. phosphatidyl choline, are added to pure water the surface tension is reduced but in this case as the barrier is moved towards the platinum strip, the surface tension is reduced still further. This is because phospholipids are relatively water insoluble and are concentrated at the available surface. Their charged groups are water soluble and

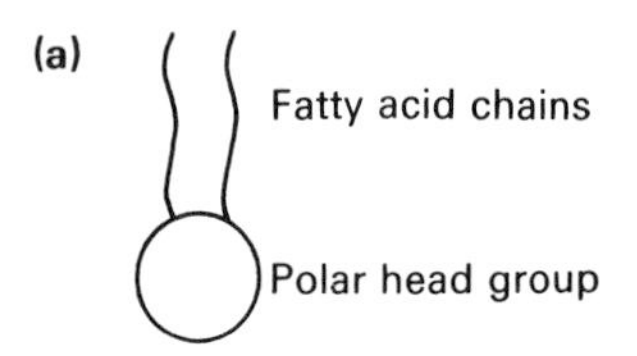

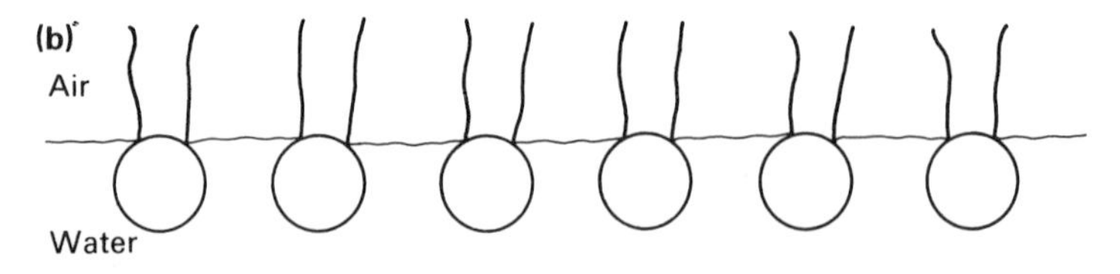

Fig. 5.9. (a) Phospholipids are surface active species; they consist essentially of a polar group and one or more fatty acid chains. (b) Because of the hydrophobic nature of the fatty acid chains, phospholipids gather at the water-air interface, and are oriented so that the polar groups make contact with the water and the fatty acid chains stick up into the air. (See also Chapter 8)

so remain in contact with the water, but their fatty acid chains are hydrophobic entities and in fact wave about in the air above the surface. As the barrier sweeps towards the strip the phospholipid molecules are forced closer and closer together and the surface tension is progressively lowered. In fact if a known amount of phospholipid is added, the surface balance will give a measure of the area per molecule when the barrier has been moved up to a position of minimum area (M in Fig. 5.10). As the

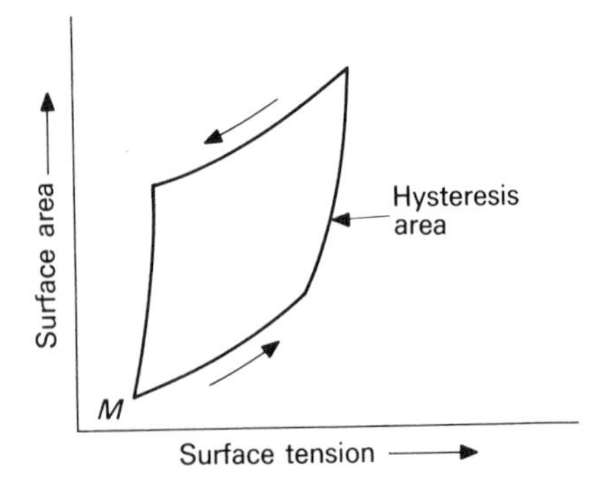

Fig. 5.10. The surface tension encloses a hysteresis curve.

energy to bring the molecules together to a certain configuration is not the same as that required to separate them, the surface tension – area curve for inward and outward sweeps of the barrier encloses a hysteresis curve (Fig. 5.10), i.e. net work has to be done on the system during each cycle.

5.6 Surface Tension and the Lung

One of the most important interfaces to terrestrial animals is the blood-epithelium-air interface in the lungs. In order to promote gaseous interchange, this interface is very large and in man it is approximately equal to that of a tennis court. In order to fit the vast lung area into the chest, the lung itself is highly invaginated (Fig. 5.11), and the tiny air spaces which are almost spherical in shape are called *alveoli.* The normal functioning of the lung we shall see demands a high concentration of certain surface active species in the walls of the alveoli.

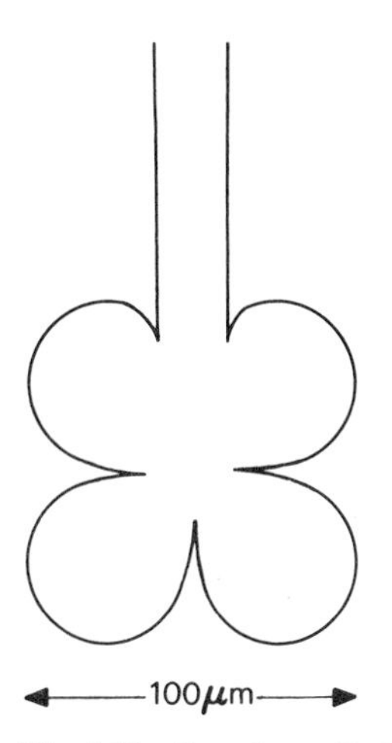

Fig. 5.11. In order to fit the vast surface area of the lung into the chest it is highly invaginated, and the figure shows a diagrammatic cross-section of an alveolus.

Karl von Neergaard, a pioneer investigator of lung mechanics, first demonstrated the important role of surface tension by some ingenious and simple experiments (Fig. 5.12). He distended lungs first with air and then with saline and found that more pressure was required to expand in air than in saline (Fig. 5.13). In the first case both surface tension and elastic forces have to be overcome whereas in the latter there are only elastic forces. Von Neergaard obtained the surface tension contribution from the difference between the two pressure-volume curves.

Von Neergaard's technique has been used to demonstrate that the surface tension contribution is much higher in a fatal respiratory disease in new born babies, called the hyaline membrane disease, than it is in the normal lung. This finding led to a search

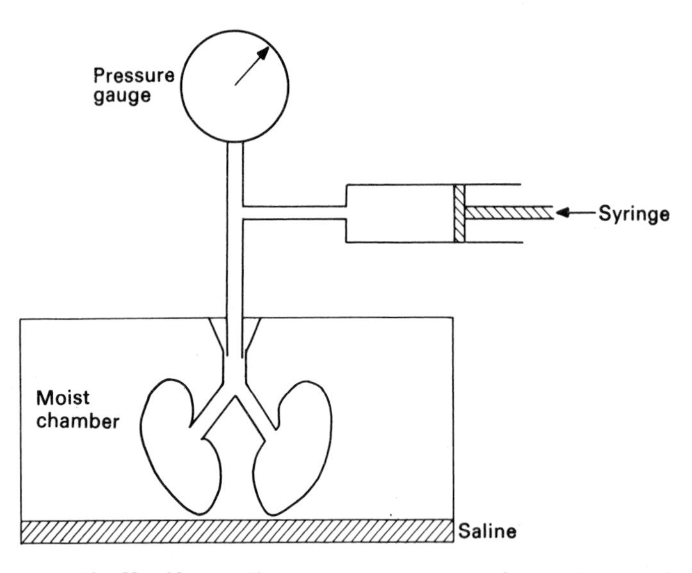

Fig. 5.12. Von Neergaard's apparatus to measure surface tension contribution to lung mechanics. The excised lungs were cannulated and connected to a syringe and pressure gauge. In this way the lung could be filled with either air or saline (After Clements, 1962). Copyright © (1962) by Scientific American Inc. All rights reserved.

for a surface active species that was reducing surface tension at the blood-air interface in the normal lung but was absent in diseased lungs. It is only relatively recently, however, with the introduction of sophisticated biochemical techniques that much progress has been made. The introduction of the surface balance (Fig. 5.8) into lung research was also required before the reason for the importance of surface tension was properly understood.

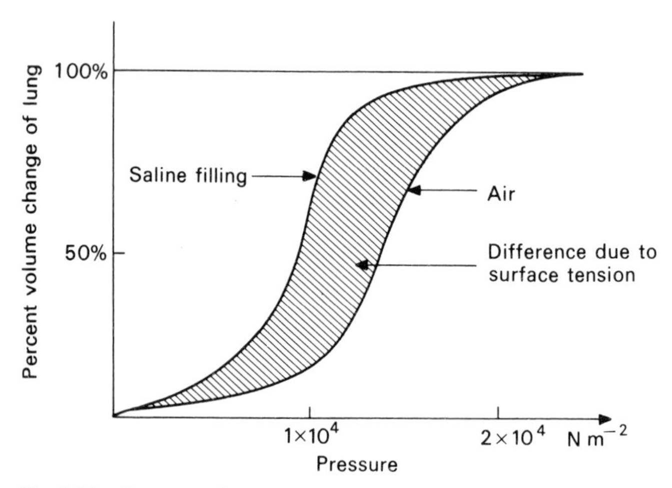

Fig. 5.13. Pressure-volume relationships when lung is filled with air or saline. (After Clements, 1962)

When an extract of normal lung is added to the water in the balance there is immediately a small but significant reduction in the surface tension (Fig. 5.14) from 70 to 40 x 10^{-3} Nm^{-1}. However there is a spectacular decrease when the barrier is moved towards the platinum strip and when a position of minimum area is reached

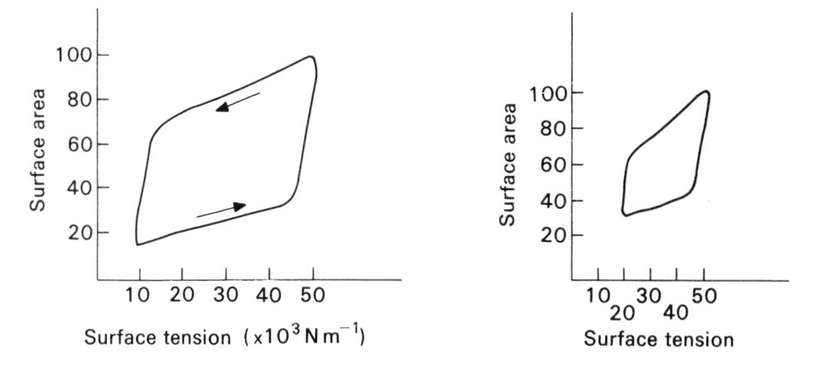

Fig. 5.14. (a) Data from lung extract in surface balance apparatus shows a hysteresis curve typical of phospholipids. At the minimum area, the surface tension is only about 8 x 10^{-3} Nm^{-1}. (b) Surface tension data from lung extract of a baby that had died from hyaloid membrane disease. At the minimum surface area the surface tension is still over 20 x 10^{-3} Nm^{-1}. (After Clements, 1962)

the surface tension has been reduced to below 10 x 10^{-3} Nm^{-1}. The shape of the area versus surface tension curve is typical of that for phospholipids (Fig. 5.10) and a relatively high concentration of phosphatidyl choline (Fig. 5.15) can be extracted from the normal lung where it seems to be anchored to the interfacial membranes by a low molecular weight protein. The surface tension from the diseased lung on the other hand does not fall below 20 x 10^{-3} Nm^{-1} (Fig. 5.14b).

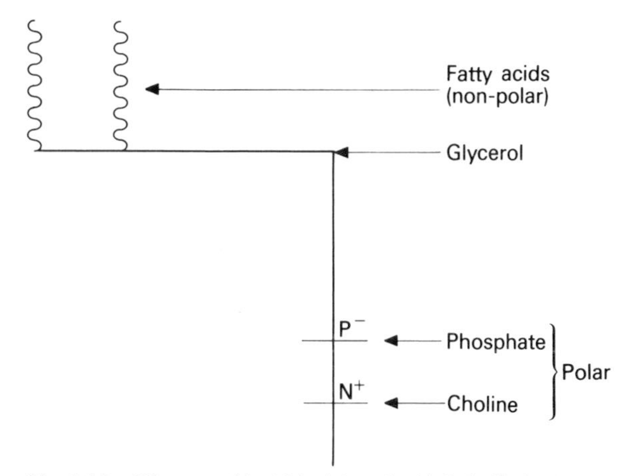

Fig. 5.15. Diagram of lecithin (phosphatidyl choline). (See also Chapter 8 for the arrangement of phospholids in membranes)

An estimation of the total possible collapse pressure on an alveolus, (of average radius 5×10^{-5} m), can be found by applying the Laplace equation 5.4

$$\Delta P = \frac{2\gamma}{r}$$

and as $\gamma = 50 \times 10^{-3}\,\mathrm{N\,m^{-1}}$ for plasma, a value of $2 \times 10^{3}\,\mathrm{N\,m^{-2}}$ is obtained for ΔP. This is in fact a considerable pressure acting to collapse the lungs and explains why the absence of a surface active agent leads to respiratory failure.

Laplace's equation also demonstrates the importance to the lung of the area dependence of surface tension. A decrease in r might be expected to lead to an increase in the collapse pressure. However, this increase is offset by a decrease in the surface tension following a compression of the surface area. Hence a relatively homogeneous distribution of pressure is brought about by the surface active species and the performance of the many alveoli, of random size, is smoothed and coordinated.

Problem 5.2 (After Jarman, 1970)
An insect called the pond skater can walk on water. It has six feet and the total length of the air-water interface at each foot has been estimated to be 1 mm.
(*a*) Assuming that the contact angle between the foot and water is sufficiently large so that the surface tension ($\gamma = 70 \times 10^{-3}\,\mathrm{N\,m^{-1}}$) acts vertically, show that this force alone can support the pond skater of mass 25×10^{-6} kg.
(*b*) Show that the skater will sink when the surface tension is lowered to $40 \times 10^{-3}\,\mathrm{N\,m^{-1}}$. (This can be rather unkindly demonstrated using detergents.)

References

Alexander R.M. (1968) *Animal Mechanics.* Sedgwick and Jackson, London.
Alexander R.M. (1971) *Size and Shape.* Arnold, London.
Clements J.A. (1962) Surface Tension in the Lungs. *Scientific American* **207** (6): 121-130.
Clements J.A., Brown E.S. & Johnson R.P. (1958) Pulmonary Surface Tension and the Mucous Lining of the Lungs; Some Theoretical Considerations. *J. App. Physiol.* **12**, 262-266.
Jarman M. (1970) *Examples in Quantitative Zoology.* Arnold, London.
Richardson M. (1969) *Translocation in Plants.* Arnold, London.
Ruch T.C. & Patton H.D. (1965) *Physiology and Biophysics.* Saunders, Philadelphia.
Sears F.W. & Zemansky M.W. (1964) *University Physics.* Addison-Wesley, Reading, Mass.
Zimmermann M.H. (1963) How Sap Moves in Trees. In *From Cell to Organism,* Freeman, San Francisco.

*Chapter 6
Sound and Ultrasonics

6.1 Introduction

The sense of hearing plays a large part in our lives; the spoken word enables us to communicate and through listening to music we both relax and derive intellectual satisfaction. Sounds also have great significance for other animals: bats use high frequency sounds for navigation and catching of food; insects have very loud courtship songs; dolphins are well-known for their ability to communicate instructions to their fellows by sound alone; and whales can in fact communicate over hundreds of miles using a code in the form of high frequency clicks.

Sound energy travels from source to receiver in the form of waves and there are certain similarities between sound and light waves. Both can be reflected or refracted at the interface between media of different compositions and both can produce *interference* and *diffraction patterns* (see Chapter 7 for an explanation of these terms). However, there are important differences.

6.2 Differences between Sound and Light Waves

(i) Sound energy is transmitted from one point to the next by vibrations of the molecules comprising the medium between the points. This implies that sound waves cannot travel *in vacuo*, an experimental fact, whereas light waves can.

(ii) Sound is a transfer of mechanical energy, whereas light is a transfer of electromagnetic energy.

(iii) Light waves are *transverse* vibrations whereas sound waves, in a gas, are *longitudinal.* In longitudinal wave motion the vibrations of the molecules in the medium are along the direction of travel of the wave. One analogy for a longitudinal wave is an arrangement of heavy balls connected by springs (Fig. 6.1a). If the first ball is set in motion by means of a horizontally oscillating force **F** (Fig. 6.1b) then a longitudinal wave will be sent out towards the right as the balls will alternatively bunch together and spread out giving regions of compression C and rarefaction R.

*Readers who have not met the concept of wave motion before should postpone reading this chapter until after Chapter 7.

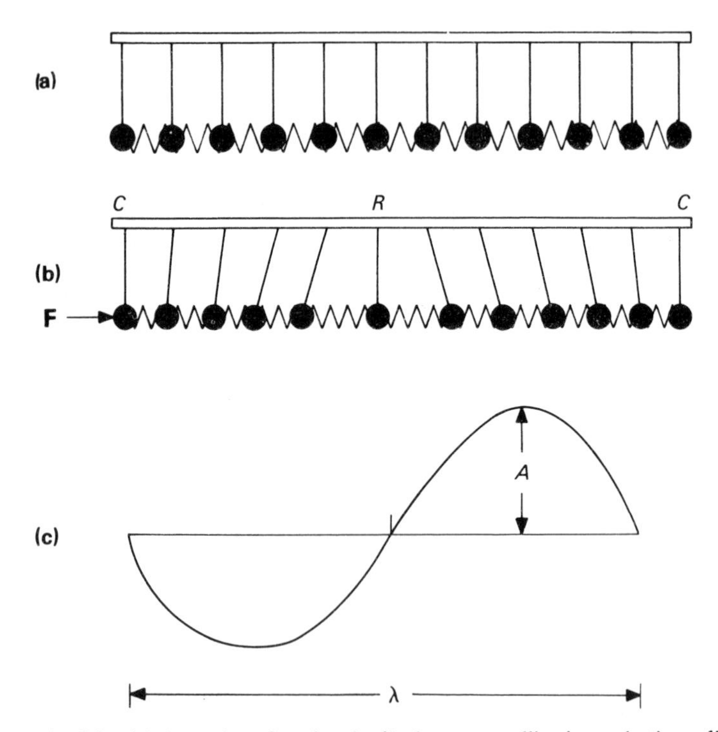

Fig. 6.1. (a) An analogy for a longitudinal wave travelling in an elastic medium. The balls represent the molecules comprising the medium and the links between them represent the compressibility. (b) An oscillating force F acting to the left will result in a longitudinal force being sent out towards the right. This consists of alternating regions of compression C and rarefaction R in the medium. (c) A is the amplitude of the disturbance. The wavelength of the disturbance is λ and this is the distance between the successive regions of compression (or rarefaction). If c is the velocity of a wave, and f is the frequency, i.e. number of cycles per second, then λ, the distance travelled by a region of compression (or rarefaction) during one cycle, equals c/f.

If the displacements of each of the balls from the horizontal are plotted, left displacements downwards, against the distance of the ball from the origin, the wave-like nature of the motion is apparent (Fig. 6.1c). Longitudinal waves can therefore be transmitted through any medium that offers elastic resistance, but the present treatment will be restricted to fluids.

(iv) The velocity of sound waves *increase* on travelling from air to water, whereas the reverse is true for light waves. The velocity of sound in air is about 330 $m s^{-1}$ whereas in water the velocity is about 1500 $m s^{-1}$. From the relationship

$$c = f\lambda \qquad 6.1$$

where c is the velocity; f the frequency; and λ the wavelength of the sound wave, which holds for longitudinal as well as transverse waves (Fig. 6.1c), it follows that the wavelength of a sound wave of a certain given frequency increases more than four fold when the wave travels from air to water.

6.3 Physical Characteristics of Sound Waves

The sounds that we hear have three characteristics: (i) *pitch,* (ii) *loudness*, and (iii) *tone quality* corresponding to the three physical quantities of *frequency, intensity,* and *waveform.*

(i) The *pitch* of a note is determined by the frequency – the more vibrations per second (Hz) of the sound source, the higher will be the pitch of the note, and in the musical scale, doubling the frequency of a note raises its pitch one *octave.*

The human ear is able to hear sounds in range 20 to 20,000 Hz and frequencies above this are said to be ultrasonic.

Any body capable of oscillating can emit sound waves and the frequency spectrum of most bodies is complex as a range of frequencies can be emitted. However, if such a body is placed within range of a tuning fork emitting a note of frequency f, then if this is one of the natural frequencies of the body, the body too will begin to vibrate and emit *sound* of the same frequency. This phenomenon is termed *resonance.*

(ii) The *intensity* or *loudness* of a sound is a measure of the energy impinging on unit area of receiver surface in unit time. The units of intensity are therefore $\mathrm{Wm^{-2}}$.

Consider the sound waves set up by a piston vibrating in an open-ended tube (Fig. 6.2). The sound wave is set up initially by the piston moving towards the right with velocity v. The piston moves through a distance vt in a time t and this sets in

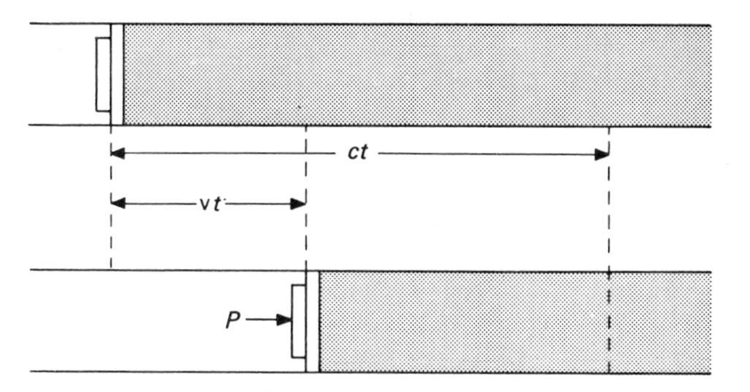

Fig. 6.2. Sound waves in a tube open at one end. A pressure P acting in the piston moves it with velocity v and this sets in motion a column of air of length ct, where c is the velocity of sound in air.

motion a column of air of length ct where c is the velocity of sound in air. P is the excess pressure, i.e. the pressure over and above the atmospheric pressure, which has to be applied to the piston to accelerate the mass of air originally contained in the column ct. The individual gas molecules in the column in fact move with a velocity v, whereas the *wavefront* of the disturbance moves with a velocity c.

As $$F = ma \qquad 1.3b$$

then $$PA = \rho\, ct\, Aa \qquad 6.2$$

where ρ is the density of air, and A is cross-sectional area of the pipe. Now as the individual gas molecules are accelerated from rest to a velocity v in time t, then

$$PA = \rho\, ct\, A\, \mathrm{v}/t \qquad 6.3$$

or $$P = \rho\, c\, \mathrm{v} \qquad 6.4$$

i.e. $$P = \text{constant} \times \mathrm{v} \qquad 6.4a$$

The constant in equation 6.4 (ρc) defines the *acoustic impedance* of the medium and note that it gives the relationship between pressure (a scalar quantity) and particle velocity (a vector).

The acoustic impedance of a *system* on the other hand, for example, the auricle of the ear, with an opening of cross-sectional area A is defined as the relationship

$$\text{Impedance} = \frac{P}{A\mathrm{v}} \quad \text{(von Békésy, 1960)} \qquad 6.4b$$

The acoustic impedance of air is 430 kg $m^{-2}s^{-1}$ and it can be shown that if there is not to be a serious energy reflection when sound travels from one medium to another, e.g. from the air to tissues of the ear, then the acoustic impedances of the two must be matched. (See Problem 6.1.)

The power carried per unit area of a column (I) is given by the force per unit area multiplied by the velocity, i.e.

$$I = \frac{F\mathrm{v}}{A} \qquad 6.5$$

$$= \rho c\, \mathrm{v}^2 \qquad 6.6$$

This is the instantaneous power or intensity and it has units Wm^{-2}. Normally we are concerned with the average intensity that is carried over one cycle of a sinusoidally varying sound wave and in this case it can be shown that

$$\text{average intensity } I_{av} = \tfrac{1}{2}\, \rho c\, \mathrm{v}^2 \qquad 6.7$$

$$= \tfrac{1}{2}\, \frac{{P_0}^2}{\rho c} \qquad 6.8$$

when the excess pressure on the piston at any time t is given by

$$P = P_0 \sin 2\pi ft \qquad 6.9$$

The intensity of the faintest sound which can just be heard is about 10^{-12} Wm^{-2} which corresponds to a pressure amplitude P_0 of about 3×10^{-5} Nm^{-2}. The loudest tolerable sound has an intensity of approximately 1 Wm^{-2} and a pressure amplitude of 30 Nm^{-2}.

Because of this wide range in intensities over which the ear operates, and because the ear can just discriminate between sounds of a certain intensity ratio whether they are loud or soft, a logarithmic rather than a linear intensity scale is used. The *intensity level B* of a sound wave is defined by the equation

$$B = \log_{10} \frac{I}{I_0} \qquad 6.10a$$

Where I_0 is an arbitrary reference intensity and is conventionally taken as the threshold of hearing, 10^{-12} Wm^{-2}. The intensity level is a dimensionless quantity and the unit is the Bel (B) or more commonly the decibel (dB) in honour of A.G. Bell. The intensity level of the threshold of hearing is taken as 0 dB and the loudest tolerable level is then 120 dB.

From equation 6.8 the decibel can also be described as

$$\text{dB} = 20 \log_{10} \frac{\text{pressure}}{\text{threshold pressure}} \qquad 6.10b$$

When the sound consists of a mixture of frequencies, as sounds generally do, then the weighting given to each frequency must be stated. There are internationally agreed weighting schemes, for instance the *A* weighting which corresponds to the sensitivity of the ear and the *D* weighting used in aircraft noise measurements. The weighting used is put in brackets after dB.

(iii) The *tone quality* corresponds to the complexity of the waveform produced by the source. A well-made tuning fork will vibrate sinusoidally with only one frequency and so produce a pure tone. However, experience shows that most vibrating bodies, in addition to the *fundamental* or lowest frequency, have *harmonics* which are frequencies that are single multiples of the fundamental. If f is the fundamental frequency, then $2f$ is called the second harmonic. The number and relative intensity of the harmonics determine the tone quality of the note. The oscillogram or graph of the waveform from the note of a violin played on the open G string is shown in Fig. 6.3a. The relative amplitudes of the various harmonics that make up the waveform are given as the *harmonic analysis* (Fig. 6.3b) and *frequency spectrum* (Fig. 6.3c). A surprising observation is that the fundamental frequency (196 Hz) is missing and this is because the body of the violin does not resonate at such a low frequency. However, even more surprising is the fact that the quality of the note is unchanged even when the fundamental is missing. It is said that the ear supplies the fundamental as this is the common *difference* tone of the upper harmonics (see following section). This fact is taken advantage of in cheap loudspeakers which give a good reproduction to the untrained ear even if they are incapable of producing low frequencies.

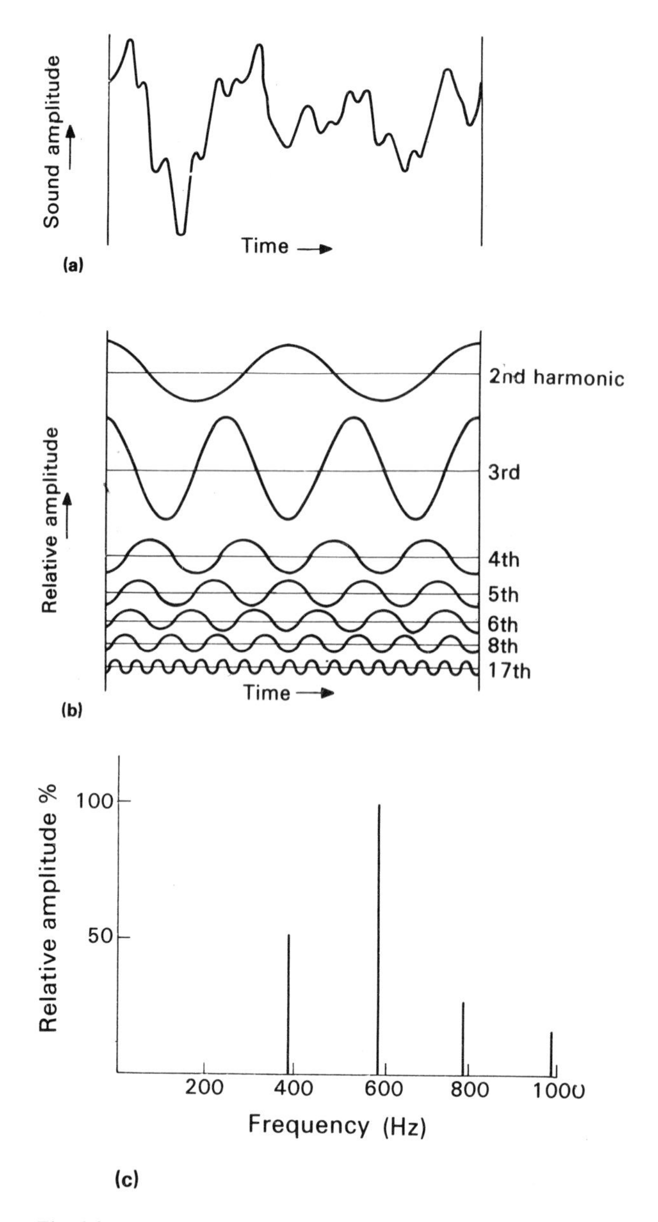

Fig. 6.3. (a) Waveform of violin tone G = 196 Hz. (b) Waveform analysed into its harmonics. (c) The frequency spectrum of a violin G string. Note that the fundamental (196 Hz) is missing.

6.4 Beats

When two tuning forks emitting different frequencies are sounded together, a periodic rise and fall in the intensity of the sound can be perceived arising from the interference of the sound waves from the two sources. The phenomenon is called *beats* and the number of beats per second is equal to the difference in frequencies of the two forks.

Consider the rotating vector representation of the waves (section 1.10) emitted from the two sources A (100 Hz frequency) and B (102 Hz frequency). Suppose that they are in phase at some arbitrary time ($t = 0$). At this time the resultant will be a maximum (Fig. 6.4a). When $t = 0.25$ s, A will have completed 25 vibrations and B 25.5; hence they will be out of phase and so a minimum will result. When $t = 0.5$ s, A will have

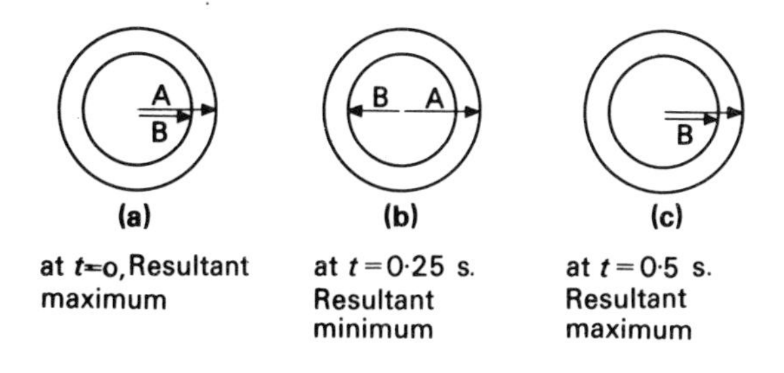

Fig. 6.4. The rotating vector representation of the waves emitted from two sources.

completed 50 vibrations and B 51. At this time the sources will be in phase and the amplitudes will add once more (Fig. 6.4c). The period between beats is thus 0.5 s and the beat frequency is 2 Hz, i.e. the difference between the two original frequencies. *Note*: interference occurs here in the waves arising from two quite different sources. Two separate light sources on the other hand cannot produce interference patterns.

6.5 Sound Production

(i) *The human voice*. Sound is produced at the vocal chords which are stimulated to vibrate by air passing out through the wind pipe from the lungs. The fundamental note of a spoken word is set up by this process and the sound is then given quality when harmonics are set up in the resonating cavities of the pharynx, mouth and nose.

An amusing change in the quality of the spoken word can be brought about by inhaling helium before speaking. As the velocity of sound in helium is higher than that in air and as the characteristic wavelength produced by the cavities does not change, the *frequency of the resonance increases* in helium. The vocal chord frequencies on the other hand, being those of stretched membranes, do not change and therefore a normally deep voice will emerge as a high-pitched squeak.

(ii) *Insect acoustics*. Insects make great use of sound in communication and in the different species a wide range of frequencies are used. As expected, the organs of hearing of a species are tuned to the frequencies emitted by members of the same species.

Just as in the well-mannered human male, the courtship sequence is initiated by sound communication. The *Drosophila* male, for example, beats a wing in the direction of his chosen female and she tunes in, probably by using velocity sensitive hearing devices made up of sensory hairs which respond to displacements of the surrounding air. These receptors are located on the anal cerci and in the antennae.

Insects make use of a very wide range of sound production and detection devices and Bennet-Clark's (1971) article is an excellent starting point for enthusiastic acoustic entomologists.

6.6 Sound Receivers

There are basically two problems which any sound receiver, including the ear, has to overcome. (i) *reflection*, and (ii) *transduction of sound energy into electrical or electrochemical energy.*

(i) Sound waves travelling in air are very efficiently reflected when they impinge on a denser medium and this loss is a result of an impedance difference between the two media. For example, 99.9% of the sound energy is reflected at an air-water interface. This figure is the same for most materials, and it means that less than 0.1% of the incident sound energy is available for conversion into electrochemical energy. Modern microphones depend on very powerful electrical amplifiers to amplify the tiny voltages produced at a relatively mis-matched transduction interface.

In the *ear*, however, there is some attempt to match up the impedance of the transducer to that of air. The first device is the ear flap or *auricle* which acts as a small ear trumpet. The sound energy is gathered from a large area and channelled into the smaller area of the *meatus* so that the forces available to set air molecules vibrating are much larger at the narrow end. This is equivalent to increasing the characteristic impedance of the narrow end, and means in fact that the impedance of the meatus is higher than that of the auricle (equation 6.4b). In this way the ear has begun to match up the impedances at the transducer interface. At the end of the meatus lies the ear drum (tympanic membrane) which is set vibrating by the sound waves. These vibrations are passed on through a system of levers called the *ossicular chain*, which is found in the middle ear. The *stapes* at the end of the chain beats against the oval window of the cochlea (Fig. 6.5). The mechanical advantage of the ossicles is approximately 2 so that the acoustic impedance of the system is once more increased because of the greater available forces and a further increment is achieved at the oval window, the area of which is only 1/20th that of the ear drum. The acoustic impedance from the air to the oval window has increased over 100 times by these devices. Waves are initiated in the cochlea canals by the vibration of the oval window; they are set up at one side of the cochlea in the perilymph of the scala vestibuli, travel through the helicotrema to the other channel, the scala tympani, and are dissipated at the round window (Figs. 6.5 and 6.6).

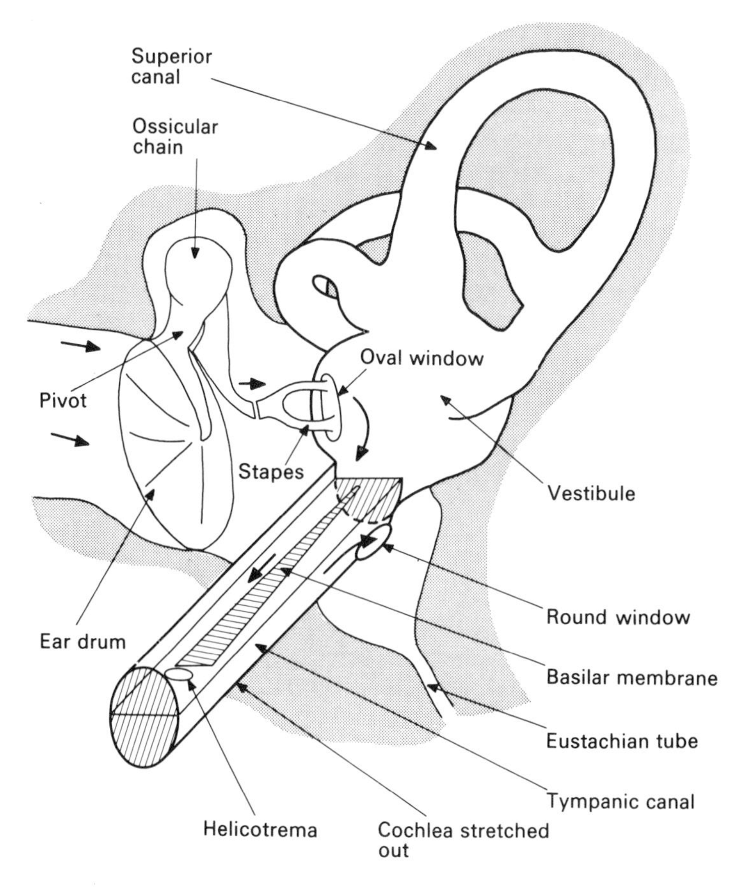

Fig. 6.5. Schematic diagram of the middle and inner ear in a mammal. The cochlea is shown uncoiled. The arrows show the displacements of fluid (air in the middle ear cavity and eustachian tube, perilymph in the vestibule and cochlea) produced by an inward movement of the ear drum. (After von Békésy, 1962)

(ii) The transduction mechanism whereby sound energy is converted into electrochemical energy at the *organ of Corti* (Fig. 6.6) is not completely understood, but one plausible hypothesis makes use of the fact that waves will travel much faster in the cochlea fluid than they will in the relatively stiff basilar membrane. Hence when a pressure wave is set up it will move down the canal and through the helicotrema before the membrane has time to move. Because the pressures above and below the membrane will not be exactly in phase, there will be a pressure difference across the membrane which will displace it and cause a bulge. The actual position of the bulge will depend on the frequency of the sound. As the basilar membrane is thin and taut near the oval window and thick and slack towards the apex, high frequency sounds will cause the bulge to appear near the window, while low frequency sounds will cause a peak near the apex. The tectorial membrane moves up and down with the basilar membrane but there is a lateral shearing stress between them which will displace the hair cells. When

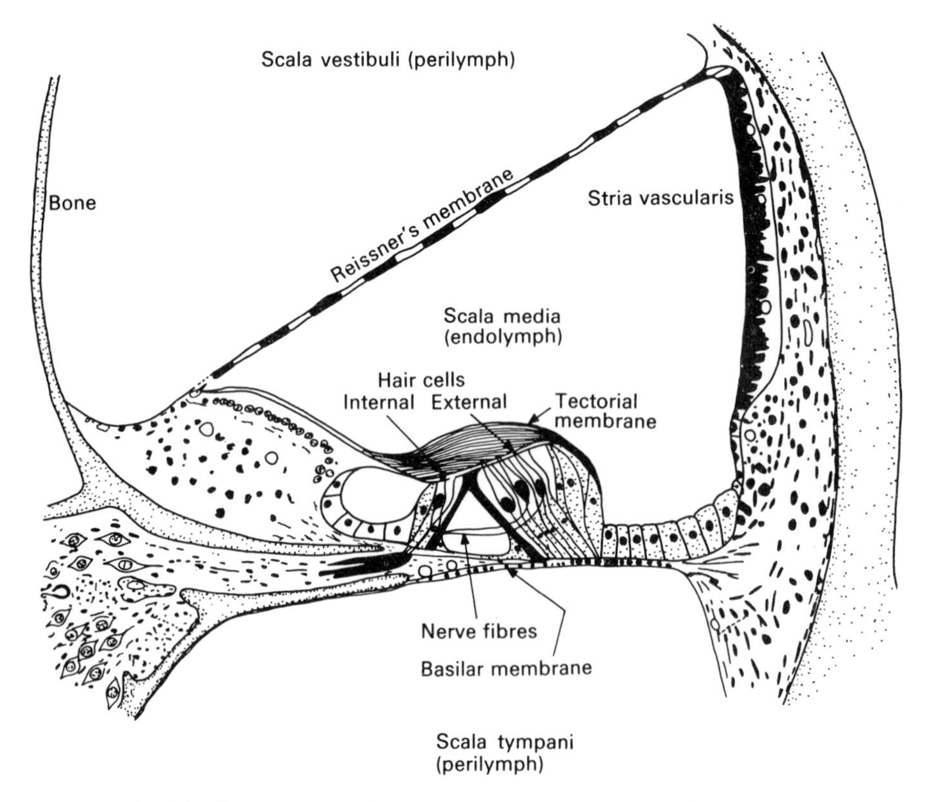

Fig. 6.6. Semi-diagrammatic transverse section of the organ of Corti in a guinea pig. (After Davis, 1953)

this displacement occurs, the nerve fibres accompanying the hair cells generate action potentials and this acoustic information will be relayed through further fibres to the brain (von Békésy, 1960).

6.7 Echo Location

The ease with which bats fly and feed at night has intrigued scientists for over two centuries. It is now known that they emit strong high frequency sounds, ultrasonics, through their nostrils and detect the echoes from objects by means of their highly developed ears. Although bats have the best known echo locating devices, the oil bird which lives in dark caves and the very interesting porpoise are only two examples of a wide range of species that possess this extra sense.

The precise mechanism involved in this extra sense is not completely understood but for the bats at least there are at present three hypotheses.

(i) The simplest means of locating the distance of an object involves timing the interval between the emission of a high frequency signal and the arrival of the echo.

While cruising, bats of the *Vespertilionidae* family emit short high frequency sound pulses which are 3 ms long and about 70 ms apart (Fig. 6.7). As the velocity of sound in air is 3.3×10^2 m s^{-1}, if the echo arrived back 60 ms after emission this would tell the bat that the object was 10 m away.

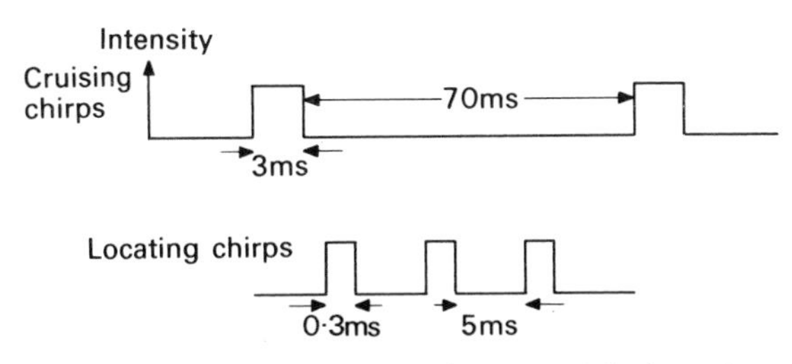

Fig. 6.7. During each chirp, the frequency falls down the scale from 110 kHz at the beginning to 40 kHz at the end, and this change in frequency during the pulse probably means that a sophisticated frequency modulating system is also involved in the range finding of the *Vespertilionidae.*

As the bat approaches an obstacle, the repetition rate of the chirps increases to about 200 per second. and the chirp width decreases to 0.3 ms. This means that the echo from an object only 50 cm away will arrive just before the next emitted pulse. The fact that the sound frequency changes during the pulse from 110 kHz at the beginning to 40 kHz at the end probably means that some sophisticated frequency modulating system is also involved in the range finding of the *Vespertilionidae.*

(ii) The second probable method of echo location depends on the ability of the bat to discriminate between echoes on an intensity basis. If the object is small compared to its distance from the bat, the intensity of the echo decreases as the fourth power of the distance of the bat from the object (problem 6.2) and by simple successive measurements of the echo intensity, a bat could perceive whether it was approaching or receding from an object.

(iii) The third method, based on a change in the pitch of the echo, is probably used by the horseshoe bats, family *Rhinolophidae*. They emit very strong well-directed beams of sound of constant frequency, and the pitch of the returning echo will depend on whether the bat and the object are flying away from or towards one another. This frequency change is called the *Doppler effect*. (See following section.)

The nostrils of the horseshoe bats have two special features that allow them to emit the required narrow beams of high intensity. The first adaptation, and one which renders these bats startlingly ugly, is a system of concave flaps round the nostrils which focus the sound waves in front of the bat. The second feature is beautifully sophisticated and involves the spacing of the nostrils emitting the sound of constant frequency (and hence wavelength, λ). The bats have evolved nostrils that are spaced exactly $\frac{1}{2}\lambda$ apart and they therefore make use of the phenomenon of *wave interference* (section 7.4). Consider a point P on the perpendicular which bisects N_1 and N_2, the nostrils of the bat (Fig. 6.8), then if the sound waves start off at N_1 and N_2 in phase,

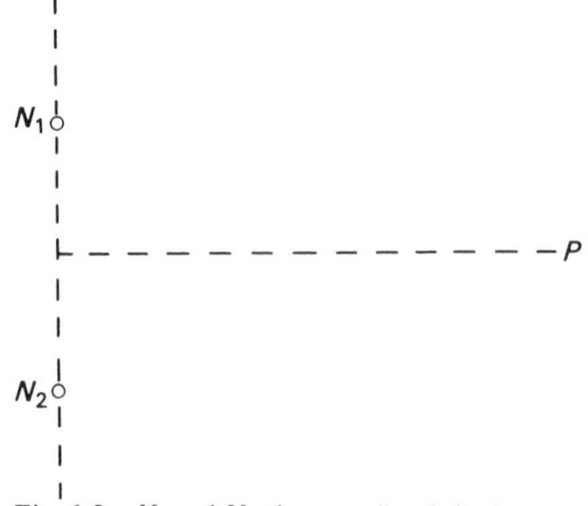

Fig. 6.8. N_1 and N_2, the nostrils of the bat, are a distance $\lambda/2$ apart. The sound energy at all points in the far field on a line drawn through N_1N_2 is zero and in fact the emitted energy is concentrated in regions in front of the bat.

they will arrive at P in phase and will give a maximum of intensity. It will also be seen that the intensity at all points on a line through N_1 and N_2 is zero as the path difference is always $\frac{1}{2}\lambda$. Hence most of the sound energy is concentrated directly in front of the nostrils.

6.8 The Doppler Effect

In the good old days of steam, there was a well known train-spotter's observation that the pitch of the train's whistle changed as it passed him in a station. As the train approaches, the observer hears a note which is higher than the true note and on passing, the pitch quickly falls to a lower note than the true pitch.

Doppler in 1842 was the first to give an explanation for this. There are three cases to consider.

(i) *Stationary source and moving observer.* Let the observer be approaching the source with a velocity v_o (Fig. 6.9a). If the observer were stationary at O, f waves would pass him in one second. However, in one second he has advanced a distance v_o to O', and he therefore hears in addition an extra vibration for each wavelength in the distance OO', i.e. he hears OO'/λ or v_o/λ extra vibrations. Hence the apparent frequency is f' where

$$f' = f + \frac{v_o}{\lambda} = f + f\frac{v_o}{c} = f\left(1 + \frac{v_o}{c}\right) \qquad 6.11$$

If the observer had been receding from the source he would have heard fewer vibrations and the frequency would be given by

$$f' = f\left(1 - \frac{v_o}{c}\right) \qquad 6.12$$

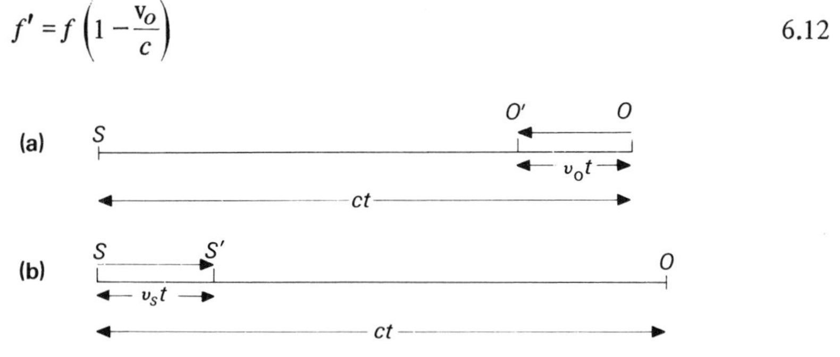

Fig. 6.9. Doppler effect, (a) Stationary source and moving observer. A source S is emitting sound waves of frequency f. The observer O is moving with velocity v_o towards S.
(b) Moving source and stationary observer. The source S approaches the observer with a velocity v_s.

(ii) *Moving source and stationary observer.* Let us assume that the source is moving towards the observer with velocity v_s. If f is the frequency of the source, then in a time t it will emit ft vibrations. At the end of t seconds suppose the first vibration has arrived at the observer, then the distance OS will be equal to ct. The source will have advanced to S' in this time and there will be ft vibrations between S' and O. With the source at rest there are s vibrations between O and S and so the wavelength of the note which the observer hears from an approaching source is less than the wavelength for the source at rest. For an approaching source

$$\lambda' = S'O/ft = \frac{c - v_s}{f} \qquad 6.13$$

but from equation 6.1

$$\lambda' = c/f'$$

hence

$$\frac{c - v_s}{f} = c/f' \qquad 6.14$$

or

$$f' = f\left(\frac{c}{c - v_s}\right)$$

If the source is receding from the observer we obtain

$$f' = f\left(\frac{c}{c + v_s}\right) \quad 6.15$$

(iii) *Both source and observer moving.* If a source of frequency f is approaching an observer with velocity v_s, relative to the ground, and the observer is also approaching the source with a velocity v_o, relative to the ground, then the apparent pitch is

$$f' = f\frac{(c + v_o)}{c - v_s} \quad 6.16$$

The algebraic sign of either numerator or denominator is changed if the direction of either v_o or v_s is reversed.

Problem 6.1

(*a*) When sound waves meet a boundary between two media, the ratio of the transmitted to incident intensities is given by the equation (Alexander, 1968)

$$I_t/I_i = 4\rho_1\, c_1\, \rho_2\, c_2/(\rho_1\, c_1 + \rho_2\, c_2)^2 \quad 6.17$$

Show that for an air-water interface only 0.1% of the incident energy is transmitted.

$\rho_{air} = 1.3 \text{ kg m}^{-3}$ $\rho_{water} = 10^3 \text{ kg m}^{-3}$

$c_{air} = 3.3 \times 10^2 \text{ ms}^{-1}$ $c_{water} = 15 \times 10^2 \text{ m s}^{-1}$

(*b*) If the acoustic impedance of the air is increased 100 fold by various devices, how much incident energy is now transmitted through the interface?

Problem 6.2

(*a*) The Inverse Square Law states that if E is the energy emitted by a point source, then the energy falling on unit area of a surface some distance r from the source is inversely proportional to r^2. Show that intensity of an echo from a point source, at the source, is inversely proportional to r^4.

(*b*) Consider two stationary objects which reflect echoes of the same intensity when the bat is 0.3 m from the smaller and 1m from the larger. Show that after the bat has flown 5×10^{-2} m towards them, the echo intensity from the smaller will be doubled, whereas that from the larger will be increased by only 20%.

Problem 6.3

(*a*) A flying bat is chasing a moving object. The bat is producing high intensity sounds, listening to the echoes, and is measuring the time between the echoes as it approaches the object. If α is the fractional increase in echo intensity from one echo to the next and τ seconds is the time between echoes, show that the bat has to fly for t seconds after the arrival of the second echo before catching up with the object, where t is given by

$$t = \frac{T}{(1+\alpha)^{\frac{1}{4}} - 1} \quad \text{seconds} \qquad 6.18$$

You may consider that the bat and object are flying with constant velocities.

(*b*) If there is a 20% increase in echo intensity with 250 ms between echoes, show that the bat has to travel for 5 s before it reaches the object.

Problem 6.4

A bat flies straight towards a wall at a speed of 10 ms^{-1} while emitting a steady note of frequency 42×10^3 Hz. What frequencies does it hear? ($c = 3.3 \times 10^2$ ms^{-1}.)

References

Ackerman E. (1962) *Biophysical Science.* Prentice-Hall, London.

Aidley D.J. (1969) Echo Intensity in Range Estimation by Bats. *Nature* **224,** 1330-1331.

Aidley D.J. (1971) *The Physiology of Excitable Cells.* Cambridge University Press.

Alexander R.M. (1968) *Animal Mechanics.* Sedgwick and Jackson, London.

Bennet-Clark H.C. (1971) Acoustics of Insect Song. *Nature* **234,** 255-259.

Davis H. (1953) Acoustic Trauma in the Guinea Pig. *J. Acous. Soc. Am.* **25,** 1180–9.

Greenewalt C.H. (1968) *Bird Song: Acoustics and Physiology.* Smithsonian Institute Press, Washington.

Griffin D.R. (1958) *Listening in the Dark.* Yale, Newhaven, U.S.A.

Mendenhall C.E., Eve A.S., Keys D.A. & Sutton R.M. (1950) *College Physics.* Heath Boston.

Sears F.W. & Zemansky M.W. (1964) *University Physics.* Addison-Wesley, Reading, Mass.

Stevens S.S., Warshofsky F. & Editors of Time-Life (1970) Sound and Hearing. *Time-Life International,* Nederland.

Taylor R. (1970) *Noise.* Penguin, Middlesex.

von Békésy G. (1957) The Ear. In *From Cell to Organism* Freeman, San Francisco.

von Békésy G. (1960) *Experiments in Hearing.* McGraw-Hill, New York.

von Békésy G. (1962) The Gap between the Hearing of Internal and External Sources. *Symp. Soc. Exp. Biol.* **16,** 267–288.

Chapter 7
Optics and Microscopy

7.1 Reflection and Refraction

Historically, the first step in the scientific study of light was made by Euclid in 300 B.C. who wrote: 'Light travels in straight lines called rays'. On this is based the science of geometrical optics and briefly, the two most important laws are:

(i) *Law of Reflection* states that when a ray of light is reflected from a plane surface, the angle of incidence equals the angle of reflection (Fig. 7.1).

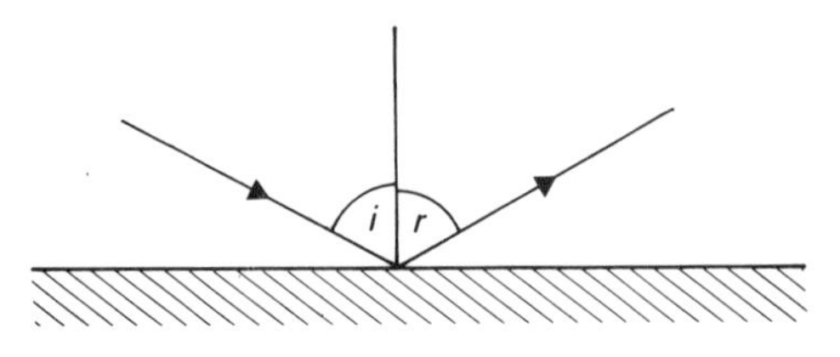

Fig. 7.1. Reflection at a plane surface. The angle of incidence i equals the angle of reflection r.

(ii) *Law of Refraction* states that when a ray of light passes from one medium to another (Fig. 7.2) the sine of the angle of incidence bears a constant ratio to the sine of the angle of refraction. The ratio is known as the *refractive index* from one medium (1) to another (2) and is denoted by ${}_1\mu_2$. Usually the refractive index of a material is

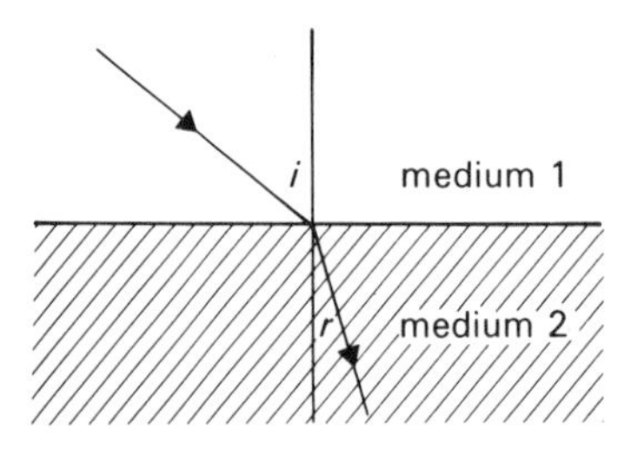

Fig. 7.2. When a ray passes into a transparent, denser medium then the angle of refraction r, is less than the angle of incidence i. The ratio $\sin i/\sin r$ is known as the refractive index of the interface (${}_1\mu_2$).

expressed relative to a vacuum, or air, which is taken as 1. In fact light is slowed down in passing from air to a medium of refractive index μ; if c is the velocity in air, then c/μ is the velocity in the medium. Newton was among the first to show that different wavelengths of light are refracted by different amounts and at an air-glass interface violet is deviated most and red least (Fig. 7.3).

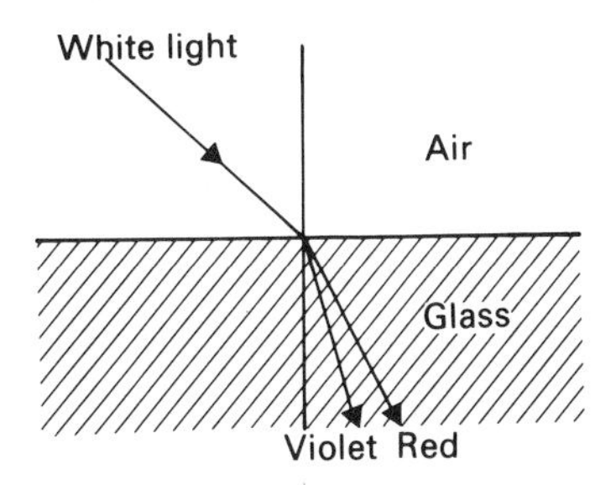

Fig. 7.3. On refraction, white light is split into its constituent colours. The violet end of the spectrum is deviated most and red least.

Curved reflecting and refracting surfaces can focus light to form images. In the case of a mirror whose surface is a part of the surface of a sphere, the equation relating the distance of the object from the mirror u, the image distance v, the radius of curvature of the mirror r, and the focal length of the mirror f is

$$\frac{1}{v}+\frac{1}{u}=\frac{1}{f}=\frac{2}{r} \qquad 7.1$$

There are several sign conventions in use and one quite good one is that distances are measured from the mirror and they are regarded as positive if they are in the same direction as that of the incident light and negative if they are in the opposite direction. A *concave* mirror focuses parallel light on to the focal point in front of it (Fig. 7.4a) while a *convex* mirror causes the rays to diverge so that they appear to come from a point behind it (Fig. 7.4b). Using the above sign convention, the focal length of a

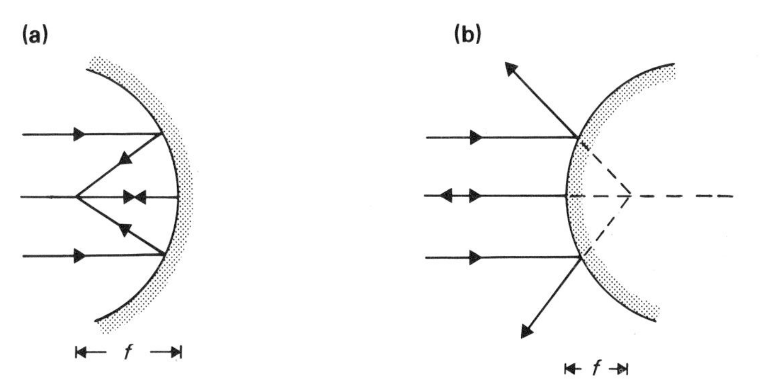

Fig. 7.4. (a) Concave mirror; the image formed is real. (b) Convex mirror; a virtual image is formed.

concave mirror is negative, while that of a convex mirror is positive. The magnification of a mirror system is given by the relationship

$$\text{image size/object size} = v/u \tag{7.2}$$

Curved refracting media also focus light (Fig. 7.5), and the lensless eye is an example of such a system. The equation to apply in this case is

$$\frac{\mu_2}{v} - \frac{\mu_1}{u} = \frac{\mu_2 - \mu_1}{r} \tag{7.3}$$

With the sign convention as before, μ_1 and μ_2 are the refractive indices of medium 1 and medium 2 with respect to air and r is the radius of curvature of the interface.

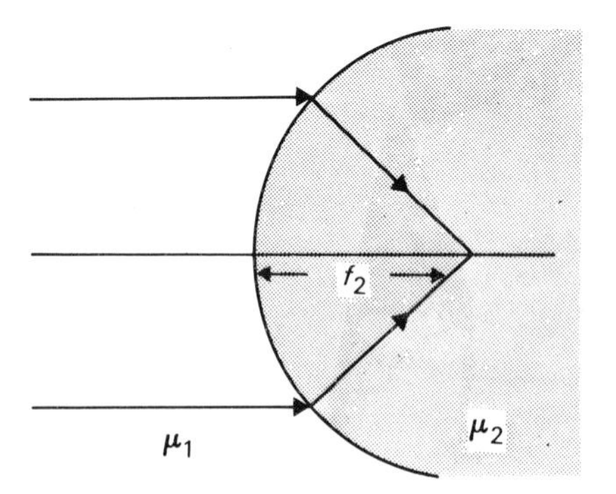

Fig. 7.5. The object is at infinity in the medium of refractive index μ_1, while the image is at the focal point in μ_2.

The magnification of such a system is

$$\frac{\text{image size}}{\text{object size}} = \frac{v\mu_1}{u\mu_2} \tag{7.4}$$

Lenses (Fig. 7.6) consist of a homogeneous refracting medium bounded by surfaces of equal curvature. The equation for thin lenses is

$$\frac{1}{v} - \frac{1}{u} = \frac{1}{f} \tag{7.5}$$

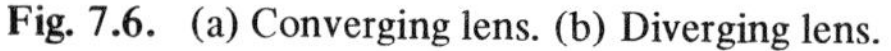

Fig. 7.6. (a) Converging lens. (b) Diverging lens.

The *power* of a lens is defined as the reciprocal of the focal length and the units are *dioptres* with units m^{-1}. The lens of the eye approximates only very crudely to a thin lens, as not only is it quite thick, but its refractive index is non-homogeneous and increases towards the centre of the lens.

The magnification of a thin lens is given by

$$\text{image size/object size} = v/u \qquad 7.6$$

There are two types of lens (Fig. 7.6), namely converging and diverging, and they have respectively positive and negative focal lengths. In order to solve problems involving lenses, the same sign convention as that used for mirrors can be adopted.

7.2 The Eye as an Optical System

The eye (Fig. 7.7) can be compared with a camera. It has a refracting system consisting of the cornea and lens which forms inverted images of viewed objects on the back surface of the retina. Except for the transparent cornea, a tough opaque coating called the sclera covers the eye and inside this is the dark pigmented layer, the choroid, which prevents stray light from being scattered around the eye. There is a blind spot in the eye where the optic nerve leaves.

In terrestrial animals the main refraction occurs at the cornea and the lens is only a fine adjusting device. (For fish is is not the case and the lens is very round, has a very high refractive index, and accommodation is achieved by moving the lens back and forth.) In man, the eye accommodates by changing the shape of the lens.

The ciliary muscles form a ring round the lens and under normal conditions, with the eye relaxed, they keep the front surface of the lens fairly flat. The lens then focuses parallel light on to the retina. If near objects are to be viewed then the focal length of the lens is decreased by the ciliary muscles contracting and the front lens surface becomes more curved. The focal length, however, cannot be decreased indefinitely and the eye cannot focus on the retina images of objects closer than a certain distance, the least distance of distinct vision, d, normally about 0.25 m for a young adult. The furthest point that can be seen clearly is the far point of the eye D.

If D and d are the distances in metres from the eye of the far and near points respectively, then $1/d - 1/D$ is called the *amplitude of accommodation* of the eye and is measured in dioptres. This decreases with age as a result of a loss in elasticity of the lens. For a ten year old child, $D = \infty$ and $d = 7 \times 10^{-2}$ m and so the amplitude of accommodation of the eye is 14 dioptres.

In a camera, light is focused on to light-sensitive, silver bromide emulsion, film. In the eye, the light-sensitive surface takes the form of photopigments located in basically two types of photoreceptor cells, namely *rods* and *cones*. The receptor cells record the arrival of light energy at the retina, this information is processed to some extent in retinal neurones, and the information is carried for further processing to the brain by ganglion cells which leave the eye in a bundle at the blind spot. Many rod receptors feed one ganglion cell so there is a pooling of visual information with a consequent

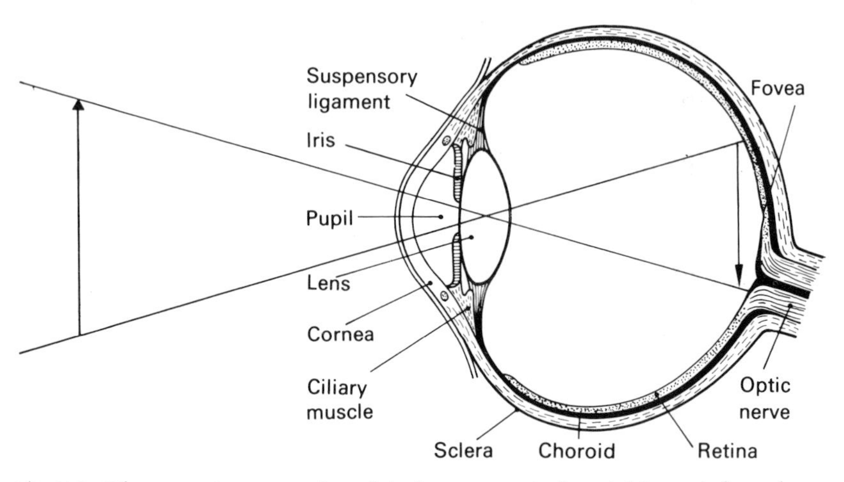

Fig. 7.7. Diagrammatic cross section of the human eye. An inverted image is formed on the retina at the back of the eye. In terrestrial animals the main focusing device is the cornea because the main refractive index difference occurs at the corneal/air interface. In amphibian and aquatic species, however, the lens plays a greater role and in these species, it is usually harder and more spherical. Accommodation in these animals is usually achieved by moving the lens nearer, or further away from, the retina rather than by altering the shape of the lens. In all vertebrate types, however, it always seems somewhat perverse that the light has to pass through several nerve cell layers before reaching the photoreceptor cells, especially as several lower animals with a similar eye structure (e.g. squid and octopus) have managed to evolve a retinal system where the photoreceptors are at the front surface of the retina.

increase in sensitivity. However this pooling leads to a loss in *visual acuity*, i.e. in apparent sharpness and detail of the object that is being viewed. Rod cells contain only one photopigment, called *rhodopsin* and so no colour vision is possible using these receptors. On the other hand, there are three types of cones, blue, green, and yellow, each containing a different pigment, so colour vision is possible with this system. The cones are concentrated around the optic axis of the eye at the fovea and these are the receptors that are mainly used when looking straight at an object. As there is more or less one ganglion cell for each cone, visual acuity is high for the cones, but the level of sensitivity is much less than for the rods.

Apart from having the ability to switch from one type of receptor to another, the eye has further advantages as it is linked to sophisticated physiological data processing and storing devices in the retina and brain. This means that we can relate the image presented on our retina with past visual experiences. Also if we are presented with conflicting visual evidence (Fig. 7.8) we continuously scan the picture trying to make sense of it. Hence the image of Fig. 7.8 which we see in our mind continually oscillates between the two possibilities in the picture, namely either that of a young or old woman.

Fig. 7.8. When two conflicting images are presented to the eye we see alternately one or the other. In this case, either an old or young woman.

Although it is an extremely sophisticated receptor system, the eye may however suffer from certain *defects of vision.*

(i) **Short-sight** (Myopia)

A short-sighted person can see near objects distinctly, but not distant objects. The latter are focused in front of the retina as the eyeball is too long (Fig. 7.9a). If it is

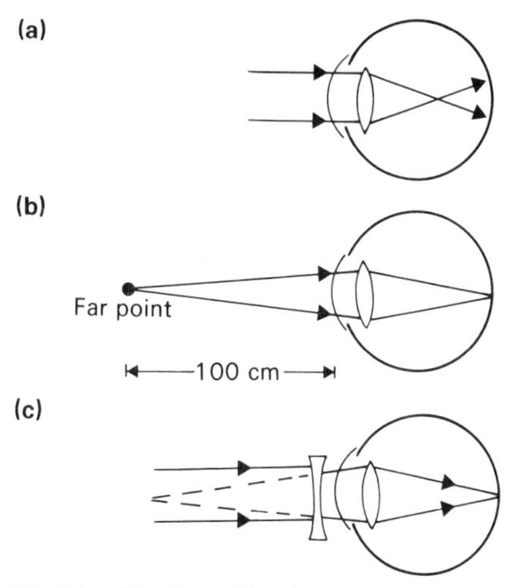

Fig. 7.9. The short-sighted eye. (a) In a short-sighted person, the eye is too long and so the light from a distant object (parallel lines) are brought to a focus in front of the retina. (b) The furthest distance from the eye that an object can be brought to a focus on the retina is called the far point. (c) In order to focus parallel light on the retina a diverging lens is used which forms a virtual image at the far point of the eye. The corneal-lens system of the eye can now focus light on the retina.

supposed that the far point is 1 m from the eye then application of equation 7.5 gives the type of spectacles to be prescribed in order that objects at infinity may be clearly seen.

$$\frac{1}{f}=\frac{1}{v}-\frac{1}{u}$$

The lens must focus the parallel rays of light at the far point in order that the corneal/lens system can focus it on the retina. When $v=-1$ m, $u=-\infty$, then $f=-1$ m.

Hence a diverging lens of 1 dioptre power must be prescribed.

(ii) Long-sight (Hypermetropia)

A long-sighted person can see distant objects distinctly but not near objects; the latter are focused behind the retina as the eye-ball is too short (Fig. 7.10a). If it is supposed that the near point is about 0.6m from the eye, then it is left as an exercise to show that converging spectacles are required to form a virtual image 0.6m distant of an object at a comfortable viewing distance 0.25m from the eye (Fig. 7.10c).

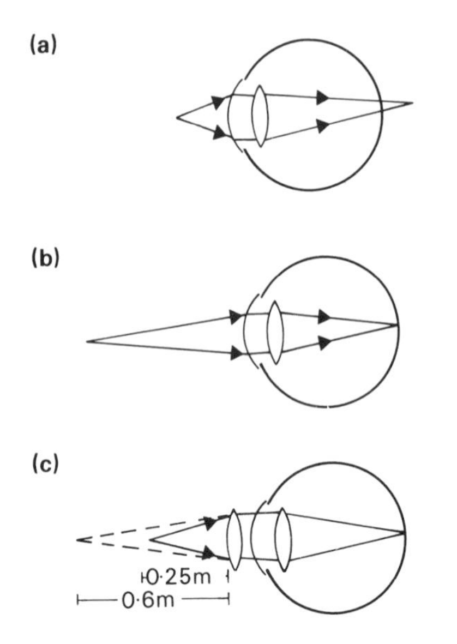

Fig. 7.10. The long-sighted eye. (a) In this case the eye-ball is too short and light from a near object is focused behind the eye.
(b) An object at the near point can, however, be clearly seen.
(c) Converging spectacles are used when viewing closer objects and these form a virtual image at the near point.

(iii) Far-sight (Presbyopia)

As the eye ages, the ciliary muscles weaken and the lens loses some of its elasticity; it therefore becomes difficult to focus near objects. To compensate, converging spectacles are required for reading. However, a short-sighted presbyope requires bifocals;

the upper half of each lens is diverging for long-distance vision and the lower half corrects for presbyopia.

(iv) Astigmatism

This occurs as a result of a lack of symmetry in the cornea and if a pattern such as that in Fig. 7.11 is viewed, then one set of lines will appear sharper than the others. This defect can be corrected by cylindrical lenses.

Fig. 7.11. Astigmatism. To a person viewing this pattern, and suffering from astigmatism, one set of lines will appear sharper than the others.

(v) Spherical Aberration

This is found on all lenses bound by spherical surfaces. The marginal portions of the lens bring rays to a shorter focus than the central region (Fig. 7.12). The image of a

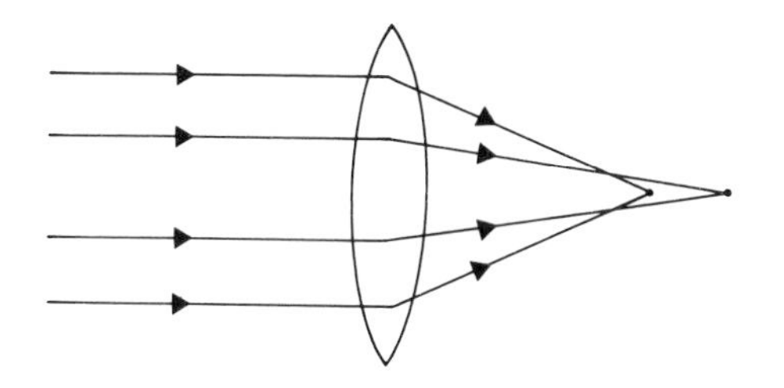

Fig. 7.12. Spherical aberration. Light from the margins are brought to a shorter focus than the central regions.

point is therefore not a point but a small blur circle. Correction by the eye is accomplished in two ways that have evolved with the eye.

(a) The cornea is flatter at its margin than at its centre.

(b) The lens is denser in the centre and hence refracts light more strongly at its core than in its outer layers.

(vi) Chromatic Aberration

All lenses made of a single material refract rays of shorter wavelength more strongly than those of longer wavelength, and so bring blue light to a shorter focus than red (Fig. 7.13). The result is that the image of a point of white light is not a white point,

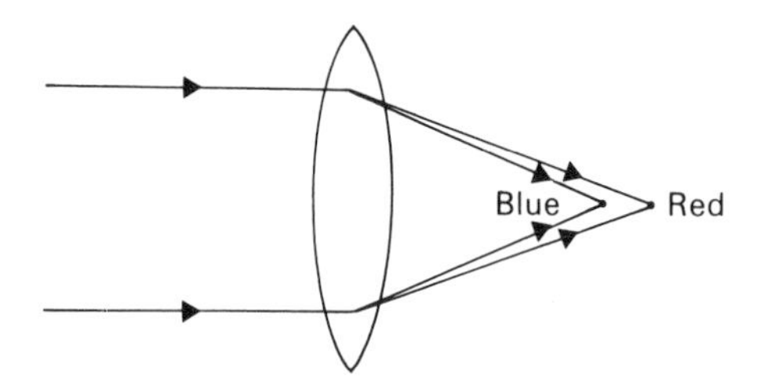

Fig. 7.13. Chromatic aberration. Shorter wavelengths are brought to a closer focal point than long wavelengths.

but a blur circle fringed with colour. Since this seriously disturbs the image, even the lenses of inexpensive cameras are corrected for chromatic aberration. The error is actually moderate between the red end of the spectrum and the blue-green, but it increases rapidly at shorter wavelengths – the blue, violet, and ultraviolet (Fig. 7.14).

As in the case of spherical aberration the eye has overcome these inherent lens defects by evolutionary processes. The first device helping the eye is the lens which acts as a colour filter. It passes the visible spectrum, but cuts off sharply at the violet end, where chromatic aberration is worst. The remaining devices are to be found in the retina itself. In 1825 Purkinje noticed that at the first light of dawn, blue objects tend to look relatively bright compared with red objects, but then they tend to look relatively dim as the morning advances. The basis of this change is in the difference in spectral sensitivity between the two types of photoreceptor cells (Fig. 7.14). Rods are maximally sensitive in the blue-green, i.e. 500 nm, whereas cones have their maximum in the green at 540 nm. Hence as one goes from dim light to bright light where pattern vision is good the sensitivity of the eye moves away from the region of the spectrum in which the chromatic aberration is large.

The third device is to be found in the fovea where the concentration of cones is highest. In man, apes and monkeys alone of all known mammals, the fovea and the region round it, called the macula lutea, is coloured yellow. The yellow pigment is xanthopyll, a carotenoid that also occurs in all green leaves.The pigment absorbs maximally in the violet and blue regions of the spectrum, just where absorption by the lens falls to low values, and hence removes further parts of the spectrum where chromatic aberration is worst.

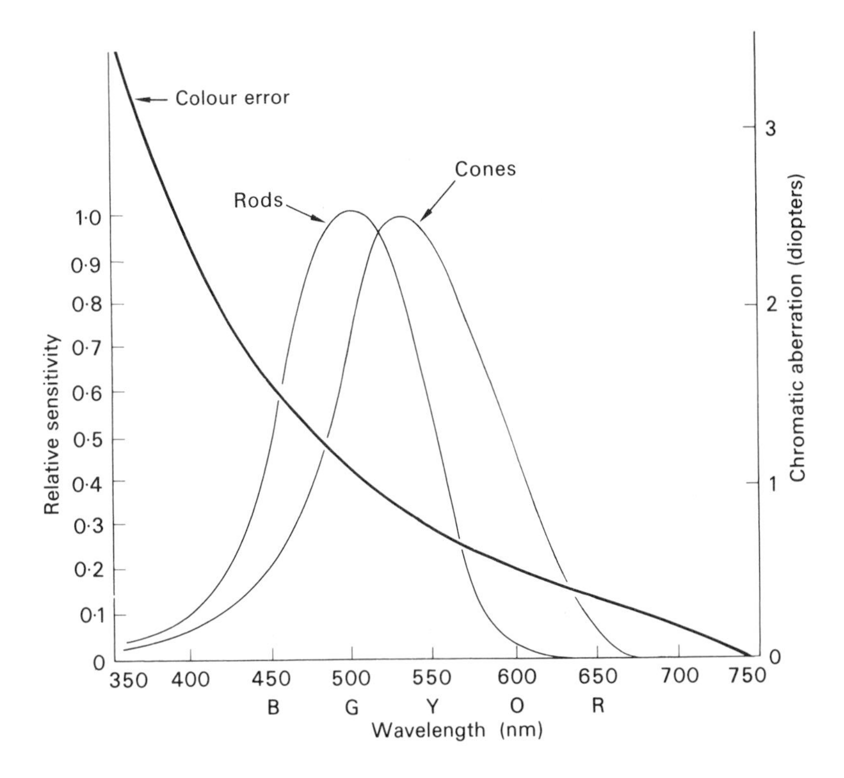

Fig. 7.14. Relative sensitivity of rods and cones and chromatic aberration as a function of wavelength. (After Wald, 1950.)

Problem 7.1 (After Jarman, 1970)

(*a*) A patient has had two cataracted lenses removed. Given that the radius of curvative of his cornea is $\frac{2}{3}$ that of the radius of his eye and the refractive index of the humours is $\frac{4}{3}$, prove that he can no longer focus parallel light on to the retina.

(*b*) Prescribe spectacles that will enable him to see distant objects.

(*c*) If he were an artist, which colour might you expect him to use more than before, and why?

7.3 Wave Theory

It has been possible so far to treat some aspects of optics without assuming any knowledge of the mechanism of the propagation of light. However, such knowledge is essential for an understanding of the principles of microscopy.

The scientific discussion of light began in the 17th century and was led on the one side by Newton who believed that light was made up of tiny corpuscles emitted from the light source like bullets from a gun. For the opposition, Huyghens maintained that light travelled in the form of waves. He also proposed that each point on a wavefront (Fig. 7.15) could act as a source of secondary wavelets and this idea is important for the understanding of the phenomenon of *light diffraction.*

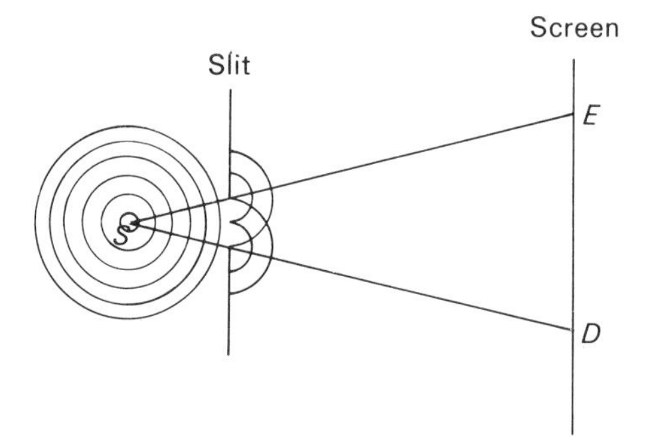

Fig. 7.15. A source S is emitting light waves. The circles represent the position of peaks at some instant in time. Light rays are perpendicular to the wavefronts and ED defines the geometrical shadow of the slit. The existence of diffraction patterns superimposed on the geometrical pattern can be explained in terms of Huyghens principle.

There were two main objections to the wave theory. It was pointed out that light can travel through the vacuum of space whereas sound waves for example cannot propagate *in vacuo* and also while sound waves may travel round corners, it appears, from the strict geometrical shadows of objects, that light cannot.

The first objection has only recently been explained in terms of the dual nature of light (Chapter 8) whereas an explanation for the second was given by Young at the end of the 18th century. He reasoned that the amount of bending of a wave depends on the wavelength and pointed out that waves of long wavelength are more easily bent. For example, if a pipe band disappears round a corner then the drum is the last sound to be heard; the higher notes of the bagpipes fade away quicker as the wavelength is smaller. Now as the wavelength of the drum sound is of the order 0.5 m and the wavelength of light is approximately 0.5×10^{-6} m, the light would be expected to bend much less and so perhaps escape notice. And indeed on close inspection light was found to bend round corners giving rise to diffraction effects at the edge of the geometrical shadow (Fig. 7.16).

Wave motion is defined as the propagation of a disturbance through a medium without the translocation of that medium. Two people each holding one end of a rope can, by moving the hand holding the rope up and down, transmit vibrational energy from one to the other without the rope moving in a horizontal direction. The waves set up in the rope vibrate perpendicularly to the resting state of the rope and

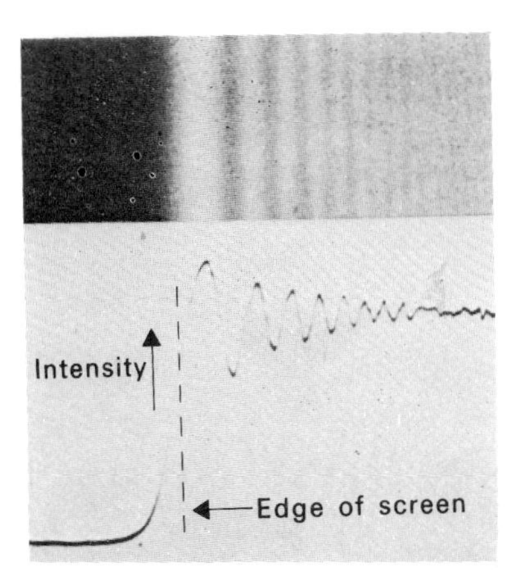

Fig. 7.16. Diffraction pattern at the edge of a slit. There is not a sudden change from darkness to light at the screen, but an oscillating intensity pattern due to diffraction (see section 7.5). (a) Photograph of intensity distribution at edge of slit. (b) Microphotometer trace of diffraction pattern. (From McKenzie, 1959.) Reproduced by permission of Cambridge University Press.

are called transverse waves. *Electromagnetic waves* (light and also radio waves fall into this class) are an example of transverse waves and this type of motion can be expressed mathematically in terms of sine waves and also rotating vectors (Fig. 7.17).

The simplest equation to represent a wave mathematically is

$$y = a \sin \theta \qquad 7.7$$

and for light waves, y could represent the electric field associated with the wave; a is the *amplitude* of the wave, i.e. peak height in Figs. 7.17 a and b, or length of the rotating vector in 7.17c. θ is the *phase* of the wave.

The time taken to complete one cycle is called the *period* and this is the time taken for the wave to travel a distance λ along the x axis (Fig. 7.17b). It is also the time that the rotating vector takes to complete a revolution, i.e. 2π radians (Fig. 7.17c). Hence we immediately have the relationship between phase angle θ, and distance x.

$$\theta/2\pi = x/\lambda \qquad 7.8$$

and so we can write another wave equation

$$y = a \sin \frac{2\pi x}{\lambda} \qquad 7.9$$

Which is equivalent to equation (7.7).

Suppose, however, that we have a wave that does not start at the origin (Fig. 7.18a) then the equation to represent this wave is

$$y_2 = a \sin (\theta - \theta_1) \qquad 7.10$$

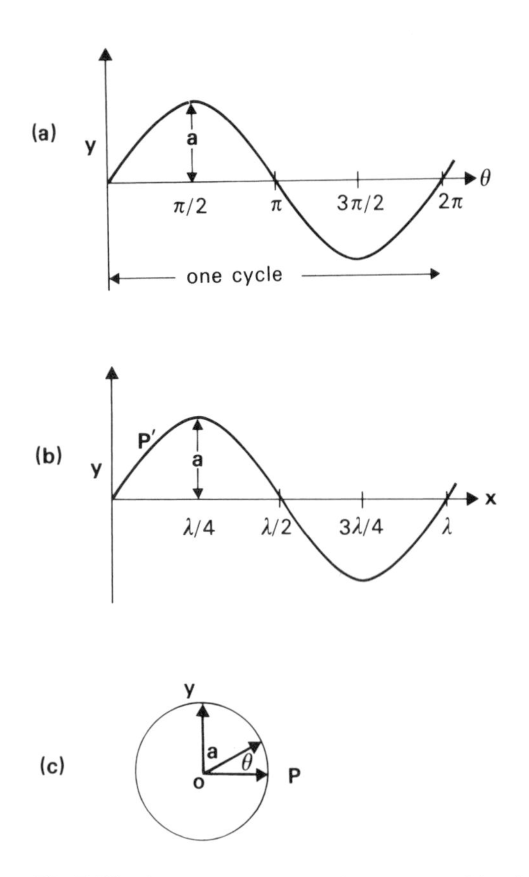

Fig. 7.17. A transverse wave can be represented by either of the equations (a) $y = a \sin \theta$ or (b) $y = a \sin 2\pi x/\lambda$. Fig. 7.16(b) in fact represents a wave of amplitude a and wavelength λ travelling along the x direction. Fig. 7.16(c) gives the rotating vector representation of wave motion. The vector **OP** represents the wave at some point in time just as P' represents the progression of the wave along the x axis at the same point in time. The magnitude of the vector represents the amplitude of the wave and the angle θ, the phase of the wave.

and this wave is said to be retarded in phase by θ_1 with respect to the first wave. The rotating vector discription of this wave is even simpler (7.18b).

If two waves are θ_1 out of phase, i.e. using the $\sin \theta$ representation of a wave, then they will also be some distance x_1 out of step using the $\sin 2\pi x/\lambda$ representation. This means that one wave will be moving along the x axis a constant distance x_1 away from the other wave. The distance x_1 is referred to as the *path difference* between the two waves and using equation (7.8) it can readily be shown that a phase difference of π radians corresponds to a path difference of $\lambda/2$ while a phase difference of 2π corresponds to a path difference of λ and so on. Of course, when two waves have a path difference of λ they will have come into step with one another and so will be in phase again.

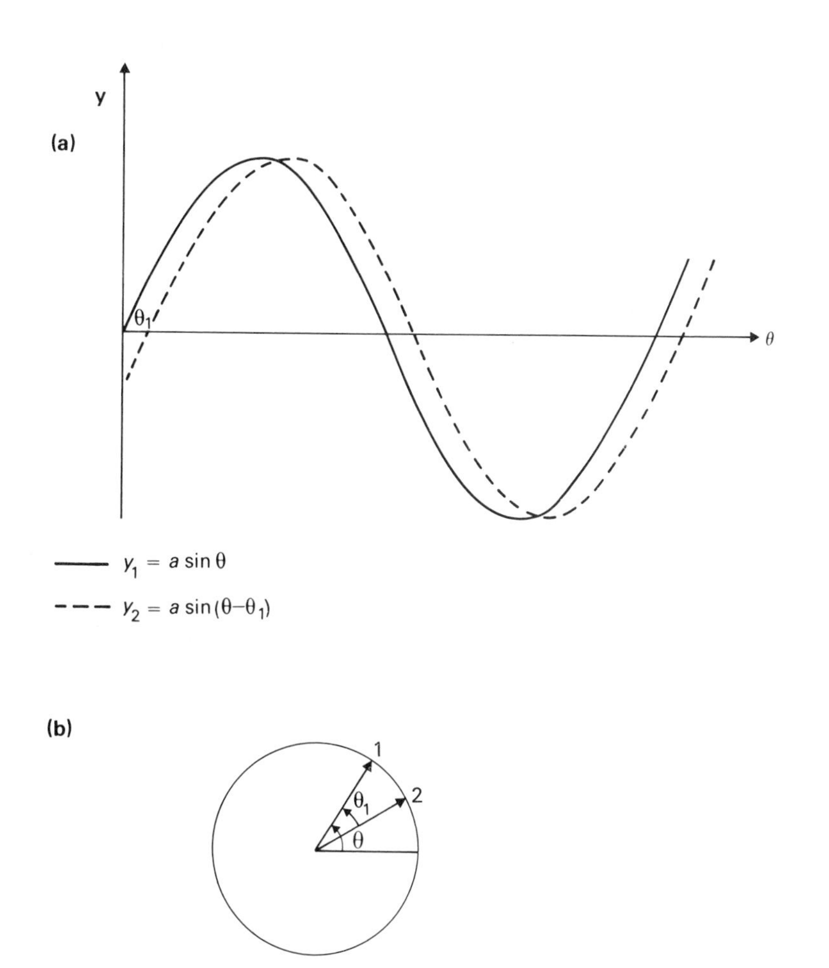

Fig. 7.18. (a) The phase difference between the waves is θ_1. (b) In the rotating vector diagram, the angle between the vectors is θ_1 (a constant) and vector 2 is said to be retarded in phase by θ_1 radians with respect to vector 1.

7.4 Interference of Light Waves

Young first suggested that if light motion was indeed wavelike in nature then it should be possible for two light waves to interact when they arrive together at the same point in space. This reasoning makes good intuitive and mathematical sense. Fig. 7.19a represents two waves of equal amplitudes and phase, and when they interact, they add to give a resultant of equal phase, but double the amplitude. Similarly when waves are π radians out of phase they interact to give a zero resultant (Fig. 7.19b). To find the resultant of two waves θ_1 out of phase is cumbersome using the wave equation

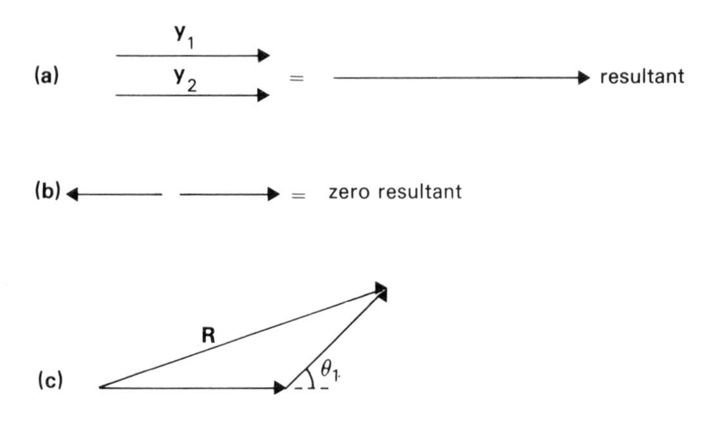

Fig. 7.19. (a) Resultant of two vectors in phase. (b) Resultant of two vectors π radians out of phase. (c) Resultant light intensity is given by R^2. Note the important point that interference patterns cannot normally be obtained from two separate sources as they will be non-coherent. This non-coherence results because the light waves emitted from most sources change phase several million times each second. This change is quite random and so two separate sources cannot be matched for interference purposes. Two coherent sources can, however, be obtained by splitting the light from one source into two separate beams, either by placing two slits in front of the lamp as in Young's experiment, or by using a half-silvered mirror (see interference microscopy, p. 107).

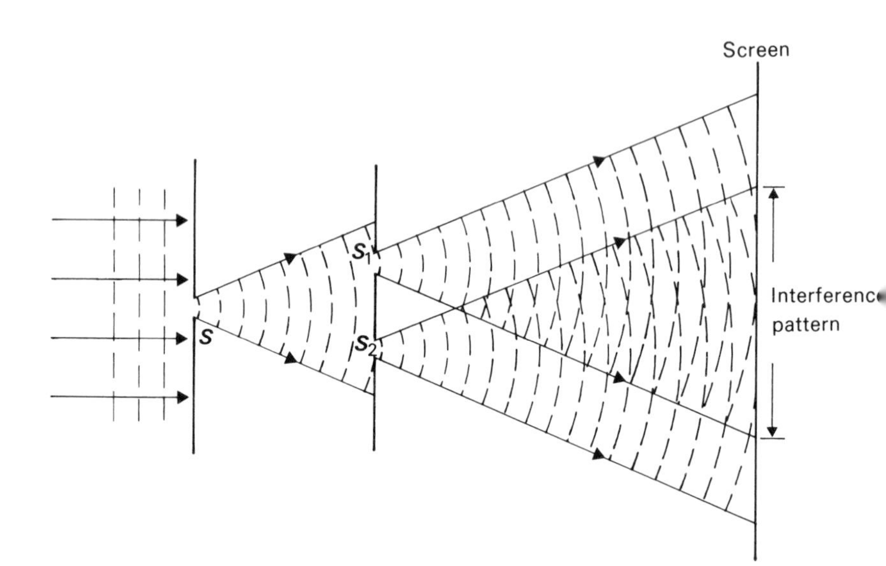

Fig. 7.20. Plane waves arriving at the first slit S give rise to secondary wavelets (diverging rays). The two slits S_1 and S_2 provide coherent sources and so an interference pattern is observed on the screen, set up by the interaction of the secondary wavelets from S_1 and S_2.

approach, but is easy when vectors are used (Fig. 7.19c), and so the vector approach will be taken in the remainder of the chapter.

Young devised an experiment to show directly the interference of light waves (Fig. 7.20). Monochromatic light, i.e. light of only one wavelength, from a narrow slit first spreads and then passes through two further slits which are close together and parallel to S. Light from these two coherent sources interfere at the screen where a series of narrow light and dark bands can be seen parallel to the slits (Fig. 7.21).

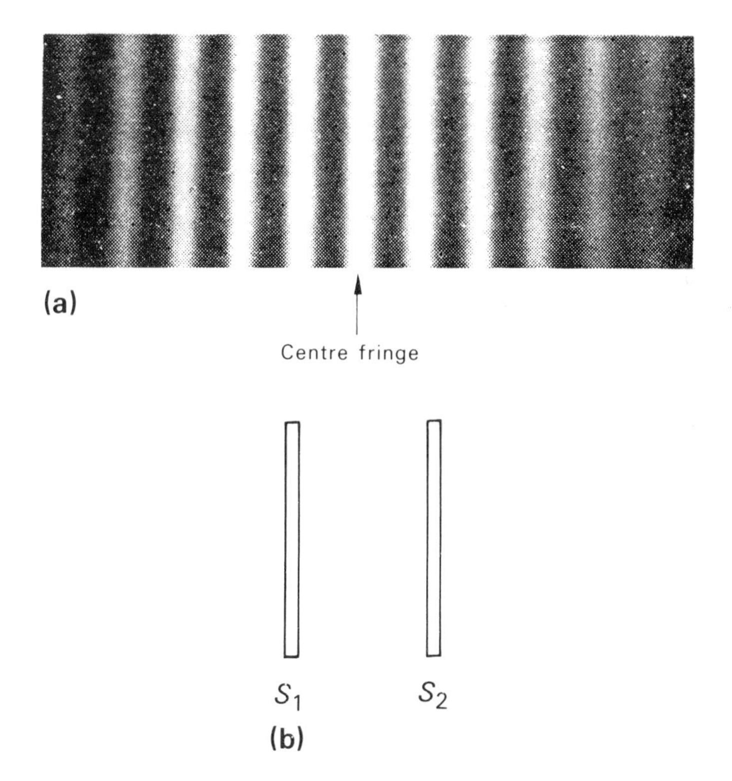

Fig. 7.21. (a) The light and dark bands seen at the screen (Fig. 7.20) are due to the interference of light waves from the slits S_1 and S_2 (b).

A simplified ray diagram of the experiment may help to show why the alternating bright and dark bands are seen on the screen (Fig. 7.22). Consider a point P on the screen. The waves from S_1 and S_2 have to travel over different distances to reach P and so although they started from S_1 and S_2 in phase they may arrive out of phase at the screen. When the path difference is zero or some integral multiple of λ (phase difference is 2π) then a bright band will be seen, and when the path difference is some odd integral multiple of $\lambda/2$ (phase difference is π) then a dark band is seen. An intermediate intensity (given by R^2 in Fig. 7.19c) will be given for all other path differences.

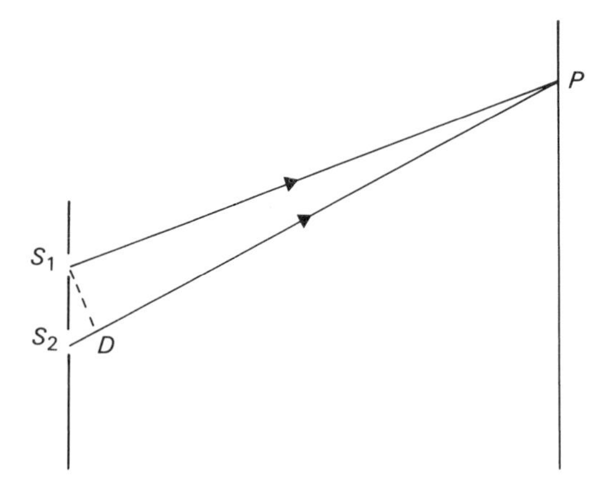

Fig. 7.22. S_2D is the path difference between the two waves and when S_2D equals $n\lambda$, then a bright band will be seen (See also Fig. 7.20).

A phase change can also be introduced between two waves (originally travelling in a medium of refractive index μ_1, say) by placing a piece of material of thickness t and of different refractive index (μ_2, where $\mu_2 > \mu_1$) in the path of one of the waves (Fig. 7.23). The velocity of the first wave (v/μ_1) is faster than the velocity of the second (v/μ_2) and so the vector representing the second is retarded to the first. How much a wave is retarded obviously depends on μ and on t. The product μt is called the *optical path length*. The difference in optical path length between the two waves is $t(\mu_2 - \mu_1)$. The phase difference introduced is $(2\pi/\lambda)\ t(\mu_2 - \mu_1)$.

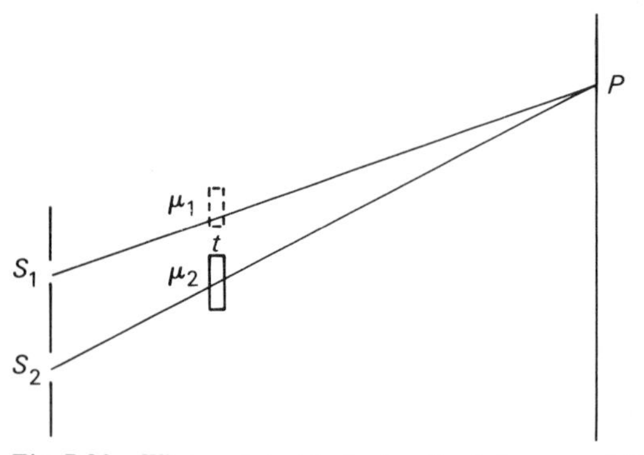

Fig. 7.23. When a material of refractive index μ_2 and thickness t is placed in the path of one of the rays then a further path difference of $t(\mu_2 - \mu_1)$ is introduced.

Note that if non-monochromatic (white) light is used in Young's experiment then coloured bands are seen.

7.5 Diffraction

Interference fringes are a special case of the more general phenomenon called diffraction which occurs whenever wavefronts are restricted by an aperture or obstacle. Instead of there being a simple geometrical shadow, a complex pattern is obtained (Fig. 7.24) which is best explained by the application of Huyghens' principle.

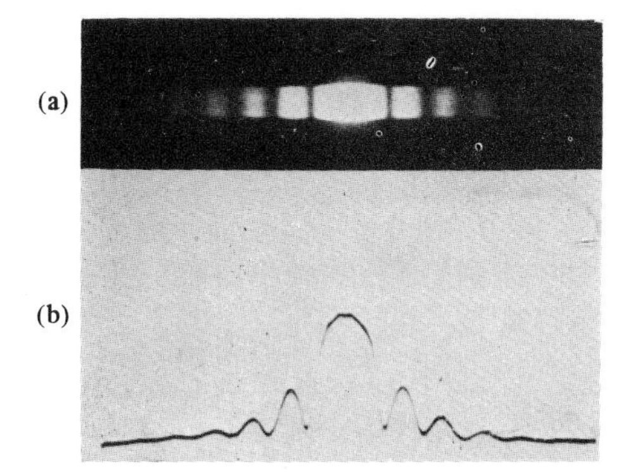

Fig. 7.24. Diffraction pattern (slit). (a) Photographic film record of Fraunhofer diffraction pattern formed from a slit of width equal to that of the central maximum. (b) Microphotometer trace of the same pattern. (From McKenzie, 1959). Reproduced by permission of Cambridge University Press.

In microscopy we are concerned primarily with Fraunhofer diffraction which takes place when parallel light impinges on the aperture or obstacle.

7.6 Fraunhofer Diffraction

Lens 1 (Fig. 7.25) produces parallel light which illuminates the slit and lens 2 collects light from the secondary wavelets in the plane of the slit and focuses them on the screen at its focal plane.

Undiffracted light is focused at *O* and light diffracted through an angle θ is collected at *P*. *SN* is drawn perpendicular to the diffracted beam. The waves arriving at *P* from *S* and *T* have a path difference *TN*. Suppose $TN = \lambda$ the wavelength of the light used, i.e. monochromatic light, then the wavelets from *S* and from the middle of the slit will have a path difference of $\lambda/2$ and hence will destructively interfere. For every point in the upper half of the slit there will be a corresponding point in the lower half for which the path difference is $\lambda/2$. Thus the first minimum or dark band will occur when $TN = \lambda$, and there will be a corresponding minimum on the other side of *O*. Minima will therefore occur when $TN = \lambda, 2\lambda, 3\lambda$, etc.

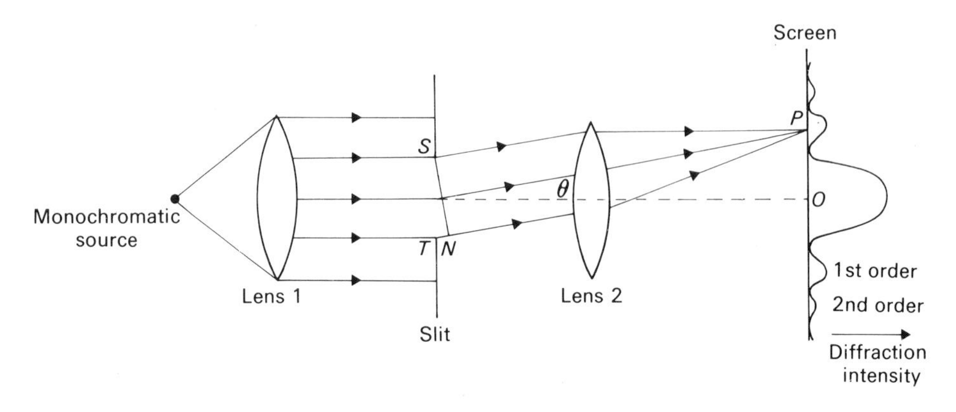

Fig. 7.25. Lens 1 produces parallel light which illuminates the slit and lens 2 collects light diffracted from the slit and focuses it at a point P on a screen placed at its focal plane. The rays from the slit represent light carried by the so-called secondary wavelets. (See also Fig. 7.15)

There will be a maximum at O as all rays travel equal distances to reach the screen. Consider the case when $TN = 3\lambda/2$, then the slit can be divided into 3 equal parts, the waves from two adjacent bands will annul, but waves from the third part will arrive to give a bright band. This bright band is called the first order maximum.

If ST, the slit width is a, then $TN = a \sin\theta$.

The conditions for minima are (see Fig. 7.26)

$$a \sin\theta = n\lambda \text{ where } n = 1,2,3.\ldots \qquad 7.11\text{a}$$

The conditions for maxima are

$$a \sin\theta = \frac{(2n+1)\lambda}{2} \text{ where } n = 1,2,3.\ldots \qquad 7.11\text{b}$$

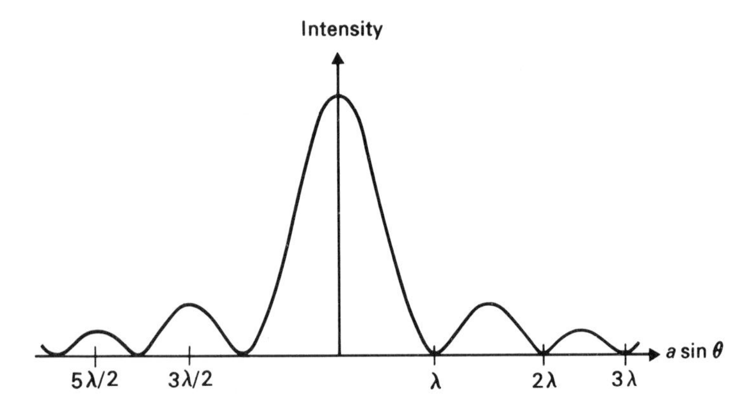

Fig. 7.26. Fraunhofer diffraction pattern. The positions of the maxima are given by equation 7.11b and their relative intensities are given by R^2 in Fig. 7.27.

The position of 1st minimum is given by

$$\sin \theta = \frac{\lambda}{a}$$

and when θ is small

$$\theta = \frac{\lambda}{a} \qquad 7.12$$

equation 7.12 shows that the smaller the slit width, the greater will be the angular separation of the minima.

This is for a rectangular slit and for a circular hole of diameter d, it can be shown that

$$\theta = \frac{1.22\lambda}{d} \qquad 7.13$$

Fig. 7.27 shows an alternative way of interpreting the diffraction pattern directly in terms of vector theory.

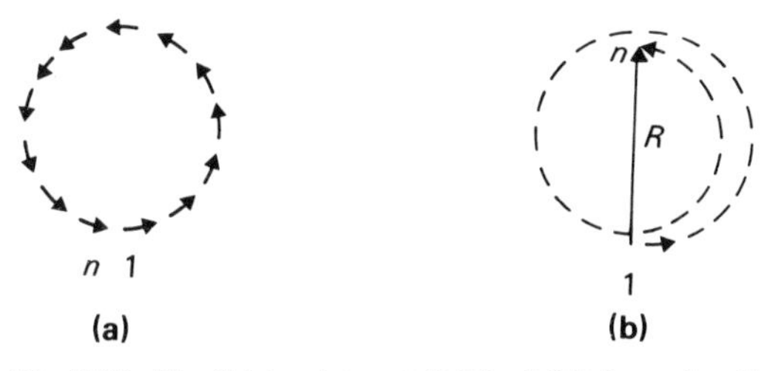

Fig. 7.27. The light arriving at P (Fig. 7.25) from the slit can be considered as n vectors set up by n secondary wavelets and the resultant can be found by adding them head to tail. (a) The phase difference between the vector representing light from the top of the slit 1 and bottom n when they arrive at the screen is 2π and so the resultant is zero. (b) When the phase difference is 3π, a maximum (1st diffraction order) is found.

7.7 Resolving Power

When a point is being viewed by a magnifying system, e.g. the telescope or the eye, only part of the wavefront from the source passes into the first or objective lens and so diffraction occurs. Instead of seeing a point image one therefore obtains a circular spot of finite extent surrounded by concentric dark and light diffraction rings. If a second point source is viewed, its image will be similar and if it is moved towards the first source a stage will be reached when the diffraction patterns will begin to overlap and the two sources will no longer be resolved. It is almost impossible to give a practical test for the ability to distinguish or resolve a diffraction pattern as being that due to two combined ones and so an arbitrary criterion, proposed by Rayleigh, has been adopted. He said that two objects would be resolved if the maximum of the diffraction pattern from one of the sources coincided with the first minimum from the other.

Light from a distant point object (Fig. 7.28) on the axis of the lens produces an image at I, and light from a distant point object off the axis forms an image I'.

The angular separation of I and I', which are the centres of the diffraction patterns, is equal to the angular separation of the distant point objects. Suppose that a circular aperture of diameter d is placed in front of the lens, e.g. the iris of the eye, then the first minimum of the diffraction pattern will occur when

$$\theta = \frac{1.22\lambda}{d} \qquad 7.13$$

and from Rayleigh's criterion, when the objects are resolved, the maximum of the second source appears here.

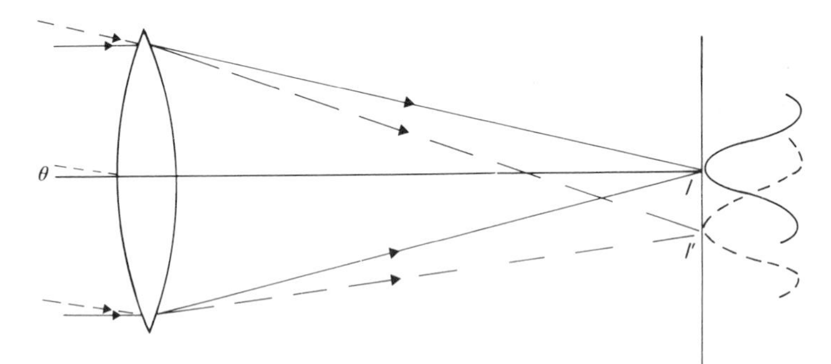

Fig. 7.28. Light from a distant object on the axis of the lines produces an image at I and light from a distant point object off the axis forms an image at I'. According to the Rayleigh criterion the two images are said to be resolved when the maximum of the diffraction pattern from one of the sources coincides with the first minimum of the other.

Applying this formula to the eye where $d = 5 \times 10^{-3}$ m and $\lambda = 5 \times 10^{-7}$ m, then $\theta = 10^{-4}$ rad. However, it is found experimentally that the acuity of the eye is never greater than 2×10^{-4} radians and so some other factor is limiting. Taking the cornea to retina distance as 2.5×10^{-2} m, then an angular separation of 2×10^{-4} radians implies that the two images are separarated by a distance of 5×10^{-6} m on the retina. Now in the centre of the fovea where visual acuity is highest, cones are about 2×10^{-6}m apart. Hence the images have to be separated by at least one unexcited cone before they are resolved, and it is this fact and not diffraction that limits *visual acuity*.

7.8 Diffraction Grating

If a diffraction grating containing many slits ruled close together is inserted in place of the single slit in the Fraunhofer diffraction experiment (Fig. 7.25) then a diffraction pattern is observed where the separation between intensity maxima is very large.

Suppose the light rays from all the slits are brought to a focus at P, (Fig. 7.25 and 7.29) then an intensity *maximum* will be observed when the angle θ is such that the phase difference between successive rays is an integral multiple of 2π, i.e. when

$$e \sin \theta = n\lambda \quad \text{(maximum, cf. for single slit)} \qquad 7.14$$

where e is the distance between slits, and n, the diffraction order, takes the values 0, 1, 2, 3 etc. As gratings normally contain over 500,000 lines per metre, i.e. $e = 1/500{,}000$ m, then the separation between the diffraction orders is very large. As $\sin\theta$ can be measured readily, the grating is used in physics to determine the wavelength of a source of light.

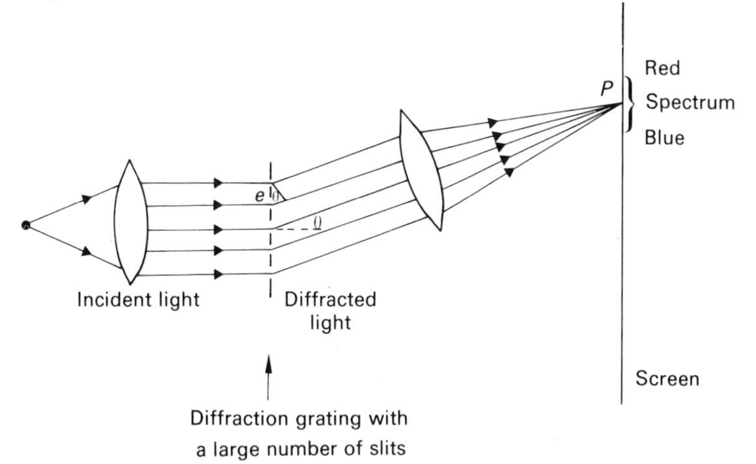

Fig. 7.29. Path difference between successive rays is $e \sin\theta$. When this path difference is λ, the rays from successive slits will constructively interfere and a maximum will be observed. When a white light source is used, a spectrum is obtained at each successive diffraction maximum (except the zeroth).

If white light is used to illuminate a grating then a *spectrum* will be produced at all maxima except the zeroth order one (in equation 7.14, $\sin\theta$ depends on λ which in this case has a wide range of values). See also section 8.7.

Any surface that has lines ruled across it at regular intervals can be used as a diffraction grating and coloured patterns can be seen in the light reflected from insect wing covers for example (Problem 7.2).

Problem 7.2 (After Jarman, 1970)
Electron micrographs of the wing cover of the beetle *Serica sericea* show that it has parallel lines across it and these are 0.8μm apart. Parallel white light falls perpendicularly on to the surface of the cover and it is viewed at an angle of 45° to the surface, what will be the colour of the wing cover?
(Blue: 450 nm; Green: 550 nm; Red: 650 nm)

7.9 Microscopy

Great advances have been made in biology with various types of microscope ranging from Hooke's original simple microscope to the modern electron microscope. In a compound microscope (Fig. 7.30) the objective forms an image of the object and this light is then imaged by the eye piece and the eye to form a virtual image beyond the objective. As in the case of the telescope, diffraction takes place in this system, and limits the resolution, only here light is scattered and diffracted by the object. For the purpose of estimating the resolving power of the system, the object can be considered as a diffraction grating. Abbé proposed the now accepted criterion for microscope resolution. He assumed that for faithful reproduction of an object by a microscope, the aperture of the objective must be wide enough to transmit at least the zeroth and both first order diffraction patterns. In fact, the more of the diffraction pattern that is lost, the poorer will be the production of the image. Suppose the object is a grating of element spacing e (Fig. 7.31) and suppose at first that the condenser lens (2) is omitted so that parallel light falls on the object,

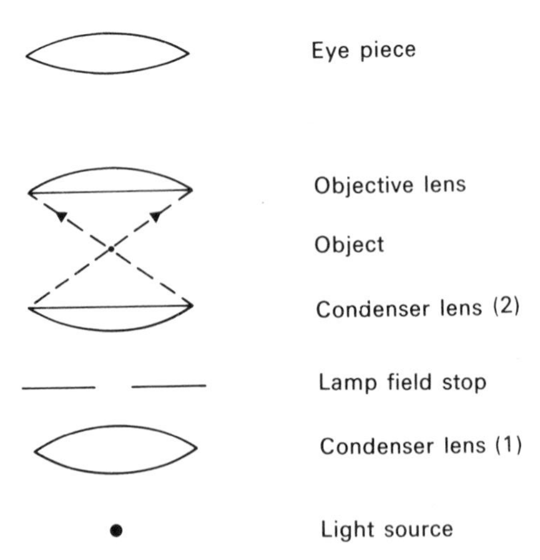

Fig. 7.30. The compound microscope (diagrammatic).

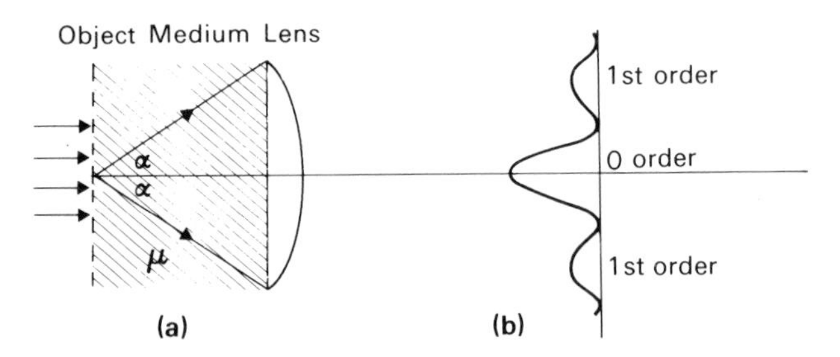

Fig. 7.31. (a) After leaving the object the diffracted light travels through a medium of refractive index μ to the objective lens of angular aperture α. (b) Diffraction pattern at objective lens. According to Abbe's criterion the aperture of the objective must be wide enough to collect at least the zeroth and first order diffraction patterns.

then in this case the position of the first order maximum is given by

$$e \sin \alpha = \lambda' \qquad 7.14$$

where λ' is the wavelength in the medium between the object and the lens, i.e.

$$e \sin \alpha = \frac{\lambda}{\mu}$$

where λ is the wavelength of the light in air, or

$$e = \frac{\lambda}{\mu \sin \alpha} \qquad 7.15$$

For e, in this case the limit of resolution, to be small, α has to be as large as possible, i.e. objective must be placed as near the object as possible. It is not always possible to illuminate with parallel light as too much light is lost from the source in illuminating a large area of the slit. A condenser lens (lens 2 in Fig. 7.30) is therefore used to focus light onto the object and this arrangement not only increases the image brightness but also the resolving power of the microscope as now the resolution equation becomes

$$e = \frac{\lambda}{2\mu \sin \alpha} \qquad 7.16$$

a fact that can only be explained by complex mathematical reasoning.
Apart from making α as large as possible, the resolution can also be improved by making λ as short as possible and μ as big as possible by using oil-immersion lenses. The ultraviolet microscope has improved resolution but suffers in that special glass has to be used in the lens construction and also the images must be recorded photographically.

In equation 7.16 $\mu \sin \alpha$ is called the *numerical aperture* and is marked on microscope objectives together with the magnification.

7.10 Interference Microscopy

The interference and phase-contrast microscopes are of great value to microscopists as they enable small objects to be viewed without the prior use of staining techniques. As there is usually only a small difference in absorption between say a cell and its bathing medium, there will be little difference in amplitude between the light which passes through the medium and that which passes through the specimen. There will, however, be a phase difference introduced because of a refractive index difference between the specimen and medium. Because the eye cannot detect differences in phase, the purpose of the interference and phase microscopes is to convert this phase difference into an amplitude difference. One way of doing this is to use round-the-square interference (Fig. 7.32).

The observed image arises from the sum of two light beams. One passes through the object and the other traverses a comparison slide by means of which the phase and amplitude of the comparison beam can be adjusted. In Fig. 7.33 the vector **OB** represents the background light and **OT** the light transmitted through the object. As

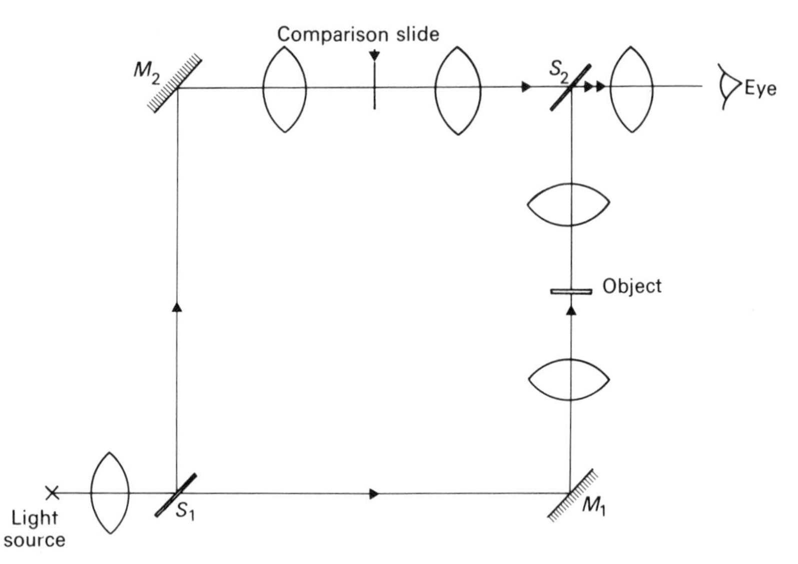

Fig. 7.32. The interference microscope. The light is split into two beams by the half-silvered mirror S_1, one half going through the object and the other half going through the comparison slide. The two beams are then recombined by S_2 and so interference is possible. M_1 and M_2 are mirrors.

the latter has a slightly higher refractive index than the suspending medium, **OT** will lag behind **OB** by a small angle θ. The phase of the light passing through the comparison slide is now adjusted until the object appears dark. The phase of the comparison beam

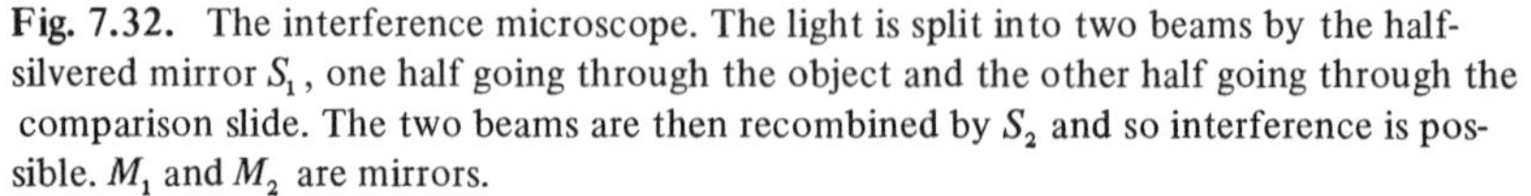

Fig. 7.33. (See text for explanation)

will have been advanced relative to **OB** by an angle $180° - \theta$ (Fig. 7.34). The object appears dark as the two vectors **OC** and **OT** are equal in amplitude but are 180° out of phase. The background does not appear dark, however, as the interaction of the two vectors representing the backgound and comparison light produces a resultant **R**. Contrast is therefore achieved and the object appears dark in a bright surround.

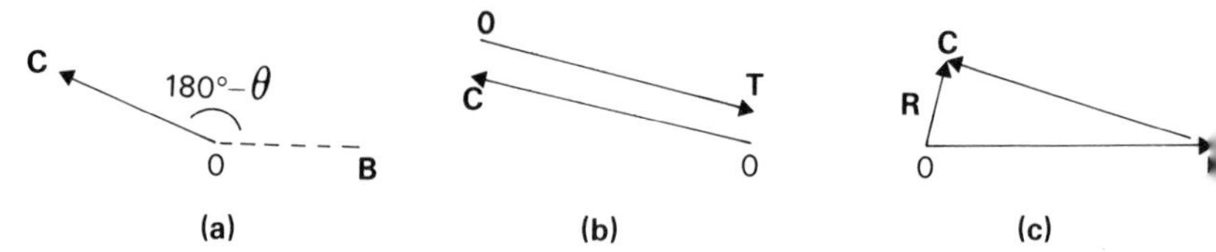

Fig. 7.34. (a) **OC** represents the amplitude and phase of the light which has passed through the comparison slide and its phase has been adjusted so that it is exactly out of phase with light transmitted through the object OT(b). The background does not appear dark, however (c) as the background vector **OB** and comparison vector are not 180° out of phase.

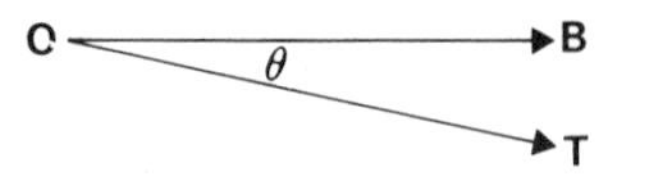

7.11 Phase Microscopy

The phase contrast microscope was invented by Zernicke and contrast in this system is achieved by interference between undiffracted light and light diffracted from the object. The phase plate (Figs. 7.35 and 7.36) and annular diaphragm lie in the focal plane of the objective and the condenser lens respectively. In the absence of an object, therefore, light which passes through the annulus falls on the etched ring on the phase plate.

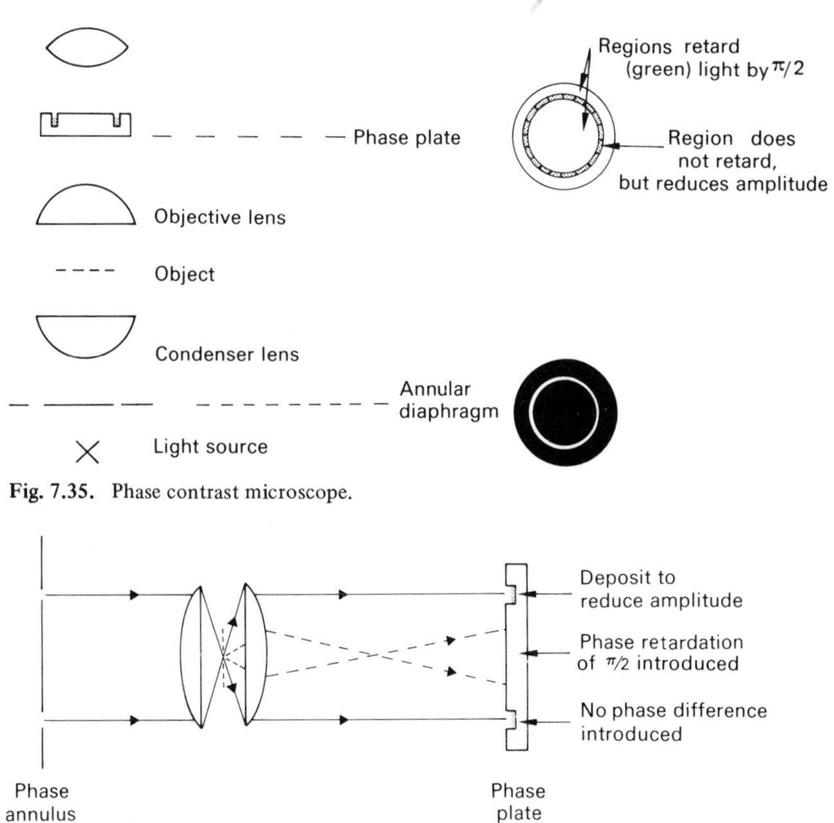

Fig. 7.35. Phase contrast microscope.

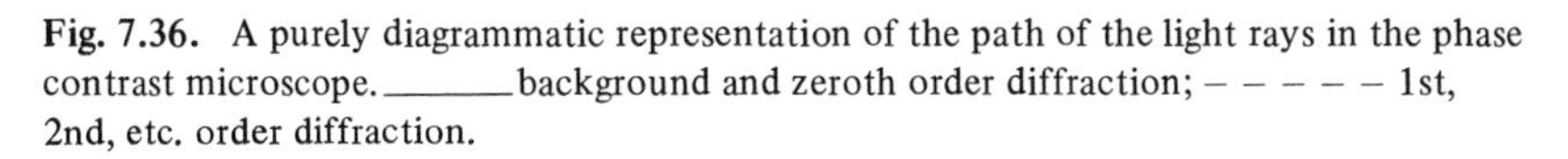

Fig. 7.36. A purely diagrammatic representation of the path of the light rays in the phase contrast microscope. ——— background and zeroth order diffraction; – – – – – 1st, 2nd, etc. order diffraction.

When an object is present, light is diffracted. The vector **OB** (Fig. 7.37) represents the background light, and **OT** the light transmitted through the object. **OT** can in fact be split into two components, **OB′**, the undiffracted light (zeroth order) which has the same phase and almost the same amplitude as the background light and **B′T** which represents the diffracted beam. When θ is small, i.e. the refractive indices of object

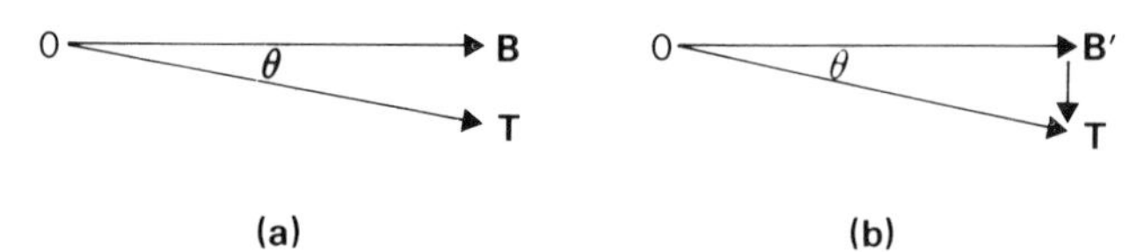

Fig. 7.37. (a) **OB** represents the background light and **OT** the light transmitted through the object. (b) **OT** can be resolved into two components, **OB′** and **B′T**; the former has the same phase and almost the same amplitude as the background light and corresponds to the zeroth order of diffraction, and **B′T** represents the 1st, 2nd and other orders and is $\pi/2$ out of phase with **OB**.

and medium are similar, then **B′T** is nearly $\pi/2$ radians out of phase with **OB′**. Undiffracted light passes through the etched region of the phase plate and is unchanged in phase but reduced in amplitude **OB**$_-$(Fig. 7.38). The diffracted light, however, strikes the plate at other points and is further retarded in phase by $\pi/2$ radians.

The zeroth and other orders combine together in the image plane to form the image. The resultant vector **I** is reduced in length compared to the background light **OB**$_-$ and the object thus appears dark on a bright background. This is called *positive phase microscopy.*

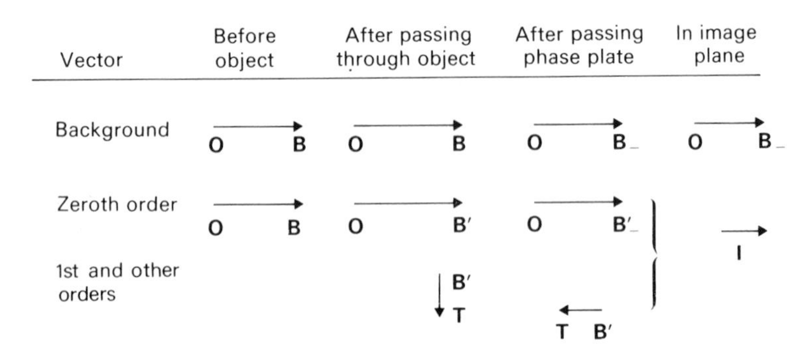

Fig. 7.38. Vector representation. B_- indicates that the background light has been reduced in amplitude on passing through the phase plate.

Note: If there were no annulus then undeviated light would also suffer a phase change and so the system would not work.

7.12 Polarization Optics

Light is an electromagnetic radiation consisting of electric **E** and magnetic **H** vectors vibrating at right angles to each other and to the direction of propagation. By convention polarization is associated with the electric vector.

In light emission from a group of atoms, the direction of vibration of the **E** vector changes about every 10^{-8} seconds and unpolarized light travelling at right angles to the plane of the paper can be represented by Fig. 7.39.

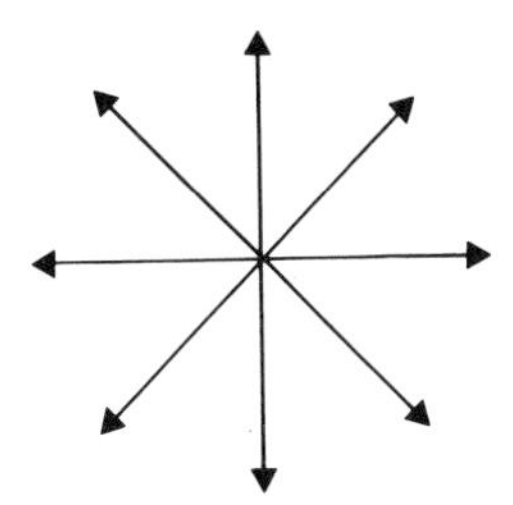

Fig. 7.39. Representation of the E vectors of non-polarized light. The light ray is travelling at right angles to the plane of the paper.

There are some special crystals that will pass light vibrating in only one direction. Tourmaline is an example of this *dichroic* class of crystals (Fig. 7.40).

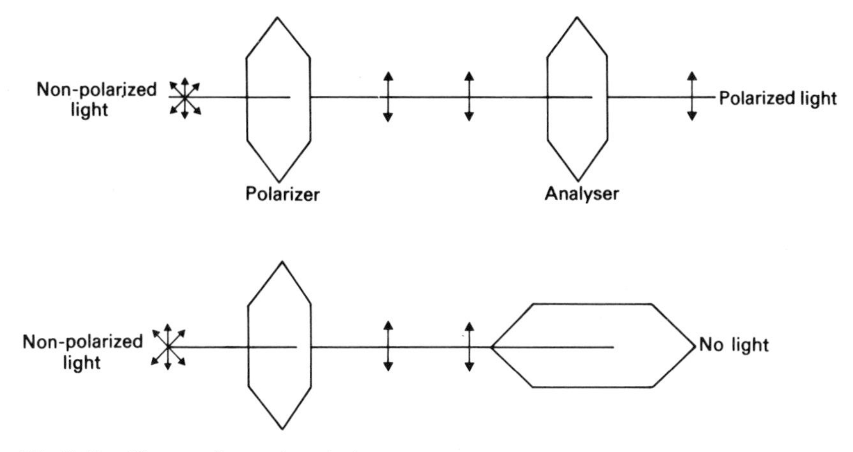

Fig. 7.40. (See text for explanation)

Only light with electric vectors vibrating parallel to the long edges of the crystal passes through. Light is said to be *plane polarized* after passing through the first crystal, the *polarizer,* and this light will be stopped by an *analyser* placed at right angles to the polarizer. As tourmalines are coloured, they are of little use in optics. Similar effects can, however, be produced by polaroid films which are not coloured. These are formed from films of polyvinyl alcohol which are first stretched to line up the polymers and then impregnated with iodine. Polaroid films are strongly dichroic and allow a small but useful range of wavelengths to pass. In the polarizing microscope, the light passes through a polaroid analyser before passing through the object and after the object the light goes through the analyser before reaching the eye. Normally the angle between polarizer and analyser ψ is under the control of the observer (Fig. 7.41).

It is only possible to get zero light intensity if the polarizer and analyser are at right angles to each other. This is because any vector can be resolved into two components (Fig. 7.41). Light is plane-polarized after passing the polarizer and if it has amplitude P, then a component $P \cos \psi$ will pass through the analyser. Hence if $\psi = 0$, i.e. the orientation of polarizer and analyser coincide, then all the polarized light gets through. If the axes are at right angles then no light emerges.

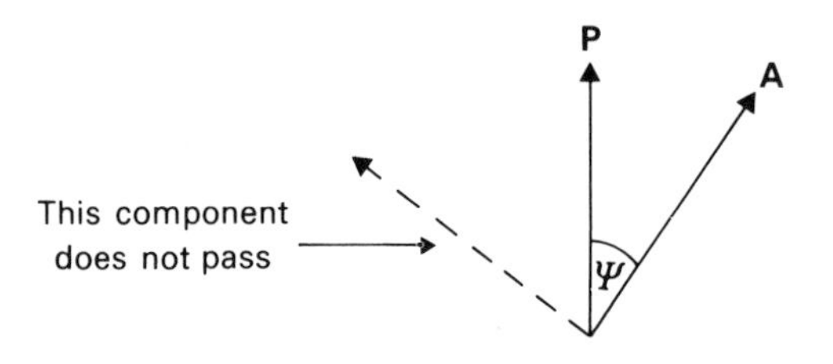

Fig. 7.41. The polarizer and analyser allow only light vibrating along the axes indicated to pass through. Only the component $P \cos \psi$ passes through both polarizer and analyser.

There are some crystals which, while not being completely dichroic, do slow down the movement of light polarized in one plane differently from that polarized in a different plane. This property is called *birefringence,* and it is shared by many arrangements of molecules and cells in living systems. We can attempt to explain this property as follows.

Consider the case of a diatomic molecule, containing atoms A and B (Fig. 7.42).

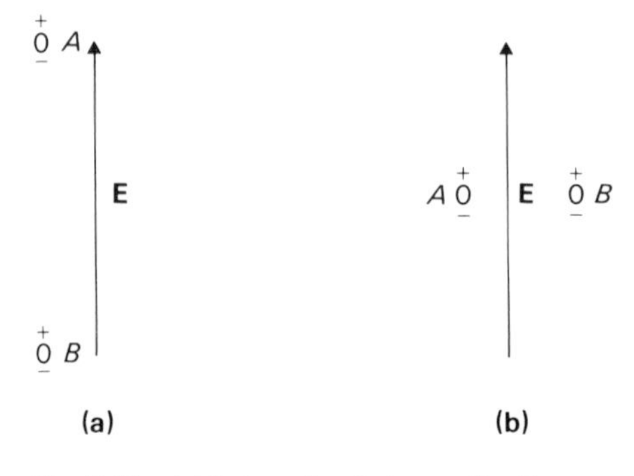

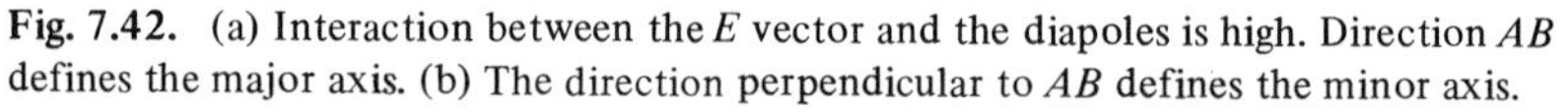

Fig. 7.42. (a) Interaction between the E vector and the diapoles is high. Direction AB defines the major axis. (b) The direction perpendicular to AB defines the minor axis.

When the light is vibrating parallel to the line joining the two atoms (a) then the polarizing effect of the light on one atom is enhanced by the induction of the other polarized atom. Dipole A enhances the polarization of dipole B and the interaction between the light and the molecule AB is high. Refractive index is a measure of this interaction. When, however, the molecule is turned through 90° (b) each dipole opposes the polarization of the other and interaction is low. In a fibre in which the chain of molecules lie parallel to each other and to the length of the fibre, it would therefore be expected that the refractive index for light vibrating parallel to the fibre length would be very much greater than for light vibrating at right angles. In a plant cell wall, for example, the direction of the cellulose chains (i.e. the direction of the *major axis* of refractive index) can be determined even in the smallest microscopically visible piece of wall. The following experiment will show this.

Set up a polarizing microscope with the polarizer and analyser at right angles, place a small piece of cell wall, e.g. from the algal cell *Nitella*, on the stage and illuminate with white light. When the stage is rotated, four positions of darkness are found showing at once that the cell wall has two refractive indices.

Suppose the long axis of the wall is at an angle ψ to the polarizer, then if light on passing the polarizer has amplitude a, the component along the direction BC will be $a \cos \psi$ and this is the amplitude of the light which will pass through the wall of thickness d. On emerging from the wall, a component of this light will pass through the analyser. The final amplitude of this light will be $a \cos \psi \sin \psi$ and the vector will be towards the right.

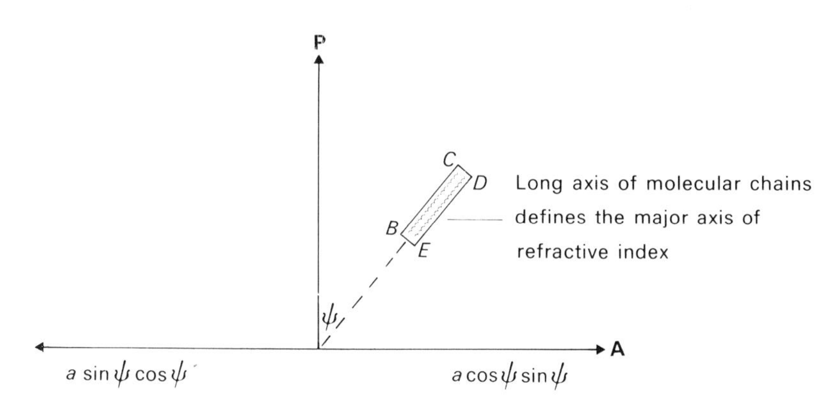

Fig. 7.43. The polarizer and analyser are set at right angles and the long axis of the cell wall is at an angle ψ to the polarizer axis.

Now remembering that light which has been plane polarized by the polarizer also has a component along the direction EB, a similar calculation shows that the amplitude which finally passes through the analyser is $a \sin \psi \cos \psi$ and the vector is towards the left.

From the figure as it stands, the two vectors have equal amplitudes, but are opposite in direction and so will give a zero of intensity on reaching the eye. However, this is only true if the cell material has equal refractive indices for light vibrating along EB and BC. In the case of the plant cell wall, the two indices are different (μ_1 in one direction and μ_2 in the other) and so a phase difference is introduced between the two vectors. The path difference is $(\mu_1 - \mu_2)d$ and the phase difference is

$(2\pi/\lambda)(\mu_1 - \mu_2)d$. When the path difference introduced is an integral number of some wavelength λ, then light of that wavelength will be missing and the specimen will appear coloured against a white background.

Suppose that the path difference introduced by the specimen is 536 nm, the wavelength of green light, then if white light is used, green will be missing and so the light will appear red. If the path difference is less than 536 nm, the wall will appear yellowish and if it is greater, it will appear green or blue. The scale which associates path difference with colour is called *Newton's colour scale* (Table 7.1) and can be used to decide which are the directions of the major and minor axes of the wall and hence the orientation of the cellulose molecules comprising the wall.

Table 7.1 *Newton's Colour Scale*

Path difference (nm)	Colour of birefringent material between crossed polaroids
0	Black
330	Yellow
536	Red
575	Violet
747	Green

In order to obtain the direction of the major axis of *Nitella,* for example, a piece of birefringent material, with its axes clearly marked, is now inserted into the polarizing microscope between the polarizer and analyser and at an angle of 45° to the polarizer. The slice of crystal normally used introduces a path difference of 536 nm between light traversing its major and minor axes and so in this case the field of view in the Microscope appears red. The colour of the plant cell wall depends on its orientation relative to the axes of the red plate.

When the long axis of the wall is parallel to the major axis of the plate, the wall appears green, which is an *addition colour* in the Newton's series. This means that the path difference introduced by the plate adds to that introduced by the wall, and hence the two major axes coincide when the long axis of the wall is parallel to the major axis of the plate. This means that the cellulose molecules run parallel to the long axis.

This conclusion can be checked by rotating the wall through 90°. It now appears yellow, which is a *subtraction colour* and from this we can conclude that as expected in this orientation it is the minor axis of the wall that is parallel to the major axis of the plate. The cell wall is said to be *positively birefringent* as the morphological long axis coincides with the major birefringent axis.

Similarly it can be shown that muscle fibres are positively birefringent, whereas myelinated nerve cells are negatively birefringent. If the slide holding the nerve is flooded with chloroform (a lipid extractant) then the sign of the birefringence changes showing that it is set up in the natural state by the radially oriented phospholipids. (See X-ray investigation of nerve, Chapter 8, p. 119).

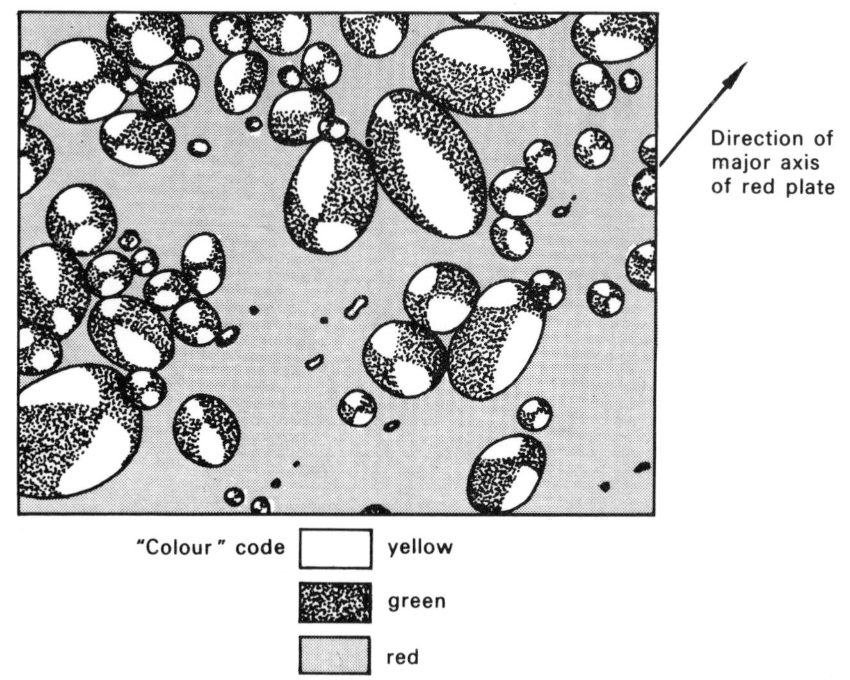

Fig. 7.44. Maltese cross appearance of starch grains. (After Preston, 1952)

One of the most beautiful examples of the use of the polarizing microscope is in the observation of a scraping from the middle of a potato. The starch grains have a Maltese Cross appearance (Fig. 7.44) and the arms of the cross parallel to the major axis of the plate show an additional colour, green, while the arms at 90° to this direction are yellow. From this we can conclude that the starch molecules are radially oriented (Fig. 7.45).

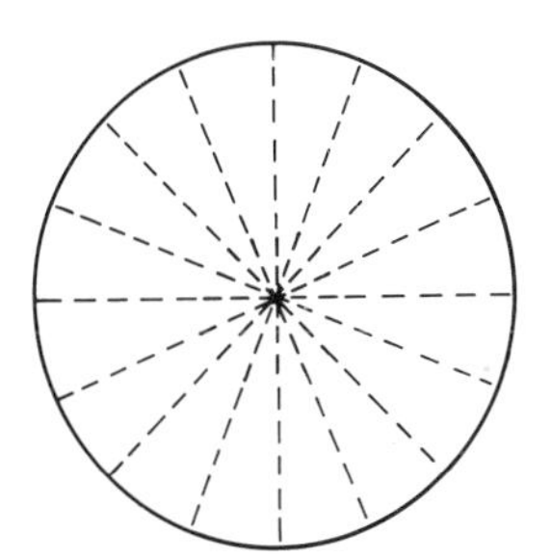

Fig. 7.45. Conclusion from the experiment shown in Fig. 7.44. The molecules comprising the starch grain are radially oriented.

References

Ackerman E. (1962) *Biophysical Science.* Prentice-Hall, London.
Epstein H.T. (1963) *Elementary Biophysics.* Selected Topics. Addison-Wesley, Reading, Mass.
Gregory R.L. (1972) *Eye and Brain.* Weidenfield and Nicholson, London.
Harrison G.R., Lord R.C. & Loofbourow J.R. (1960) *Practical Spectroscopy.* Prentice-Hall, New Jersey.
Jarman M. (1970) *Examples in Quantitative Zoology.* Arnold, London.
McKenzie A.E.E. (1959) *A Second Course of Light.* Cambridge University Press.
Preston R.D. (1952) *The Molecular Architecture of Plant Cell Walls.* Chapman & Hall, London.
Setlow R.B. & Pollard E.C. (1962) *Molecular Biophysics.* Pergamon, London.
Wald G. (1950) Eye and Camera. In *From Cell to Organism.* Freeman, San Francisco.

Chapter 8
Quantum Optics: Interaction of Energy with Matter

8.1 The Photoelectric Effect

We have seen much evidence which points to light energy being carried from point to point by wave motion. We have also been able to describe certain of the interactions of light with matter, e.g. diffraction, solely in terms of the wave theory. However, a crucial phenomenon called the *photoelectric effect* was found to be a stumbling block to a complete acceptance of the wave theory.

It was found that when light falls on a photosensitive metal plate (Fig. 8.1), electrons are emitted and can be collected at a non photosensitive anode. With a steady

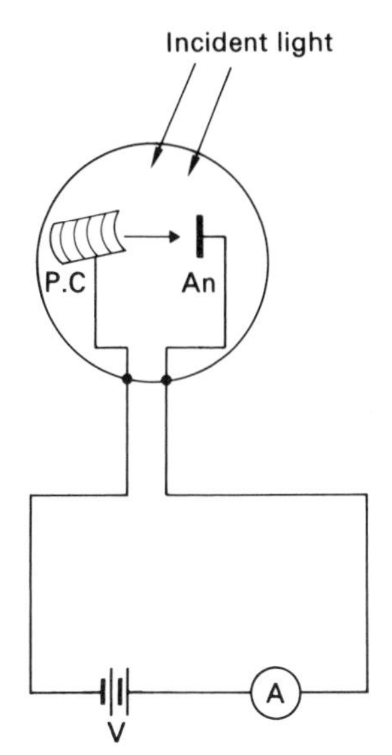

Fig. 8.1. In the dark, no current is registered on the ammeter A, but when light falls on the photocathode PC, electrons are driven off and are attracted to the anode An because of the voltage drop V between cathode and anode.

beam of light falling on the so-called photocathode, a steady current is observed to flow round the circuit. The current can be explained if it is imagined that the light energy excites the electrons so that they gain sufficient energy to leave the surface of the photocathode.

The speed, or more correctly the kinetic energy, with which the electrons leave the surface can be measured by reversing the polarity of the electrodes (see Chapter 9, p. 145 for an explanation of the electrical terms used here) so that electrons are attracted back towards the photocathode. It is found experimentally that there is a certain potential V_0 at which no current flows, and V_0 gives a measure of the energy of the electrons, i.e. the higher the energy, the greater will have to be V_0 to keep the current at zero.

From wave theory, we might expect that the higher the intensity of light (the greater the amplitude of the waves) the greater V_0 should be. However, experimentally V_0 is found to be independent of intensity. V_0 varies only as the frequency of the light used and the higher the frequency the greater is the energy of the emitted electrons.

This observation led Einstein to postulate that light energy is carried by small discrete packets, now called *quanta* or *photons*. The energy of a quantum E is given by the relationship

$$E = h\nu \tag{8.1}$$

where ν is the frequency of the light and h is *Planck's constant* which has the value 6.6×10^{-34} J s. According to the quantum theory, therefore, increasing the intensity only means increasing the *number* of quanta per second leaving the photocathode. This has no effect on the voltage required to reduce the current to zero. The higher the frequency of the light, however, the greater is the energy of the incident photon and the greater will be the kinetic energy of the electron driven off by the photon.

8.2 Light Emission

It is now well known, through the work of Lord Rutherford and others, that an atom consists of a positively charged central nucleus with electrons in orbit around it (Fig. 8.2).

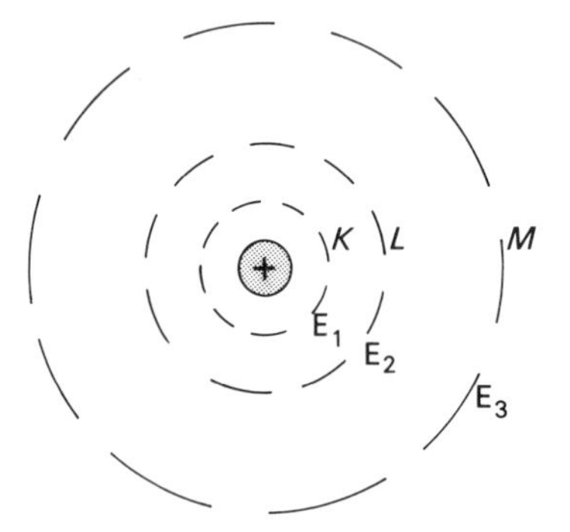

Fig. 8.2. The Coulombic interaction (Chapter 9) between the nucleus and innermost orbitals is high and so the orbitals have a high energy content $E_1 > E_2 > E_3$. When electrons from the innermost orbitals, labelled *K, L, M,* etc., are lost a very large potential energy hole is formed and so high energy photons are emitted when electrons jump down from the outer orbitals.

The basic assumptions of the Quantum Theory are:

(i) The electron orbits are such that the electrons radiate no energy while they are in them and the atom is in a low energy non-radiating state, often called the *ground state,* when all the orbitals have their prescribed number of electrons.

(ii) The excited state is achieved by the absorption of sufficient energy to remove one or more of the electrons from a given orbital; a potential energy hole is thus created and the closer the electron to the nucleus, the deeper the hole. The atom loses this excess energy in quantal form when an outer electron jumps down to fill a hole in an inner orbital. The energy of the emitted photon is in fact the difference in energies of the two orbitals involved, i.e.

$$\text{photon energy} = \Delta E$$

and as

$$c = \nu\lambda$$

where c is the velocity of light

then $$\Delta E = hc/\lambda$$

or $$\lambda = hc/\Delta E \qquad 8.2$$

When transitions are between the outermost orbitals, the energy involved is small and the photons involved are in the visible range. When innermost orbitals are involved, ΔE is large and the photons are called X-rays.

8.3 Investigation of Biological Ultrastructures by X-ray Diffraction

(i) Theoretical Background

The amount of information that can be derived from an examination of any material depends ultimately on how fine a probe is used. For example, an examination of biological tissues by the optical microscope (Chapter 7) is limited by the wavelength of visible light which is in the region of 500 nm.

Details of molecular arrangements within the tissue are not resolved as they are only 1–10 nm apart and so for these another technique has to be used where the probe is much finer. The wavelength of X-rays are in the region of 0.1 nm and so at first sight they might appear to provide an excellent probe. Unfortunately, however, no substance has yet been found that can focus X-rays in the way that a lens gathers together the diffraction patterns from an illuminated object and assembles them to form an image. In the case of X-rays we are left with the diffraction pattern which at its simplest we record on film. X-ray analyses are limited to regularly repeating structures as it is only from the constructive interference of X-rays scattered from several identical structures that a strong diffraction pattern is derived.

In order to understand these diffraction patterns some mathematical insight is required. For example, Crick, of the Crick and Watson team, realised that the cross-wise pattern on the X-ray diffraction picture from DNA (Fig. 8.3) was due to a repeating helical structure only because he had previously worked out a mathematical treatment of helix diffraction from a totally different system.

The foundation for the mathematical treatment of diffraction patterns was laid down by Sir Lawrence Bragg. He suggested that instead of considering that each atom acts as a scattering point for the X-rays, the planes of atoms that make up a repeating structure should be considered as *reflecting units* for the X-rays. The X-ray diffraction

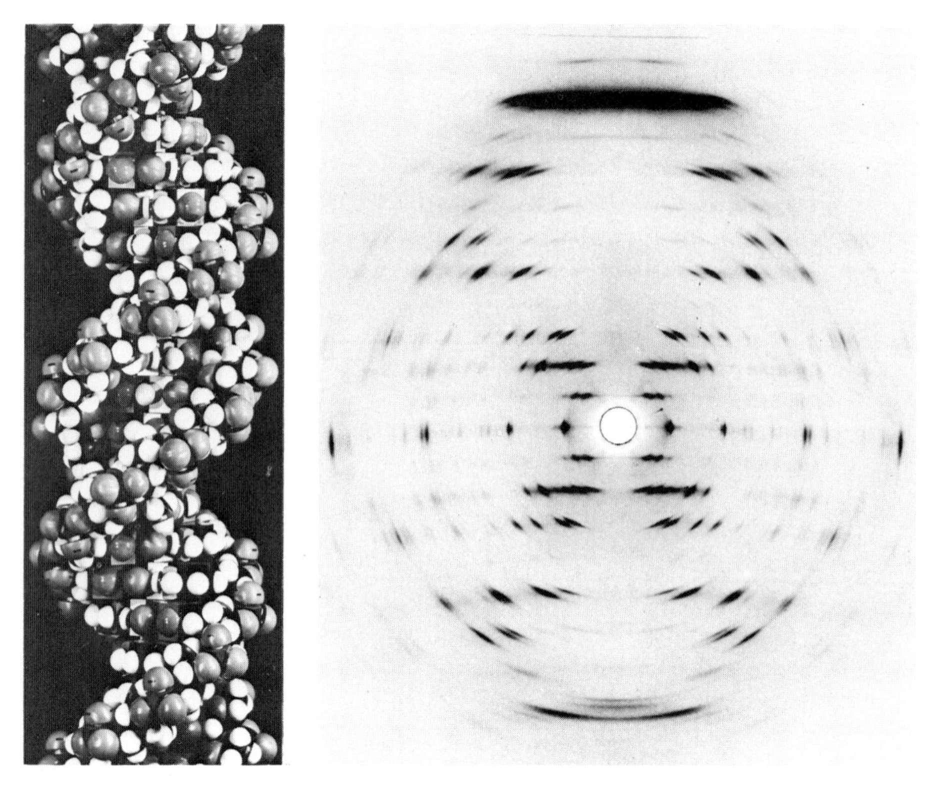

Fig. 8.3. Molecular model of DNA together with the diffraction pattern obtained from DNA crystals. Note the helical arrangement of the molecules and the cross-wise pattern of the dark spots on the diffraction picture.
(Photographs kindly provided by Professor M. H. F. Wilkins)

pattern would then be made up of all the rays reflected from the different planes. The NaCl crystal for example consists of a cubic array of sodium and chloride ions and Fig. 8.4 is a two-dimensional representation of this crystal. The X-rays can be considered as being reflected from identical planes. The path difference introduced between ray 1 and ray 2 will be $AB + BC$ and if θ is the glancing angle of the X-rays

$$\text{path difference} = 2\,d \sin\theta$$

where d is the interplanar spacing. For constructive interference

$$2\,d \sin\theta = n\lambda \tag{8.3}$$

where $n = 0, 1, 2, 3, \ldots$ is the diffraction order. Equation 8.3 is called *Bragg's equation* and is extremely important in X-ray analysis.

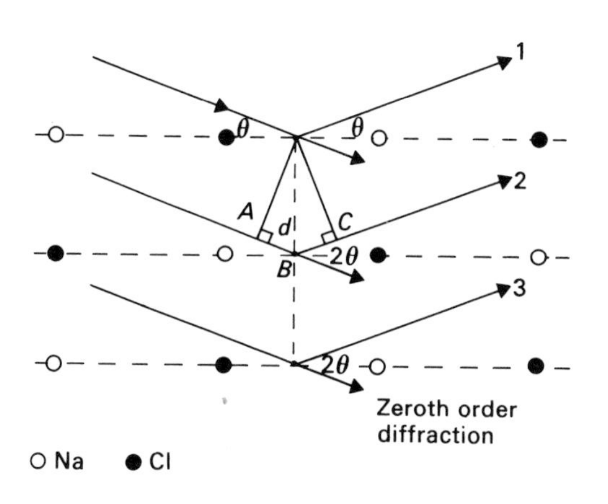

Fig. 8.4. Representation of the scattering of X-rays from repeating units (planes distance d apart) in a sodium chloride crystal. Rays 1 and 2 are reflected from identical planes and when they interfere at the plane of the film a diffraction pattern is produced. If the path difference between 1 and 2 is exactly equal to λ, the wavelength of the X-rays used, a first order diffraction spot is produced. Note that the angle between the incident and reflected rays is 2θ.

Now simpler patterns than that derived from DNA can be obtained from regularly repeating structures that have a so-called one-dimensional symmetry, e.g. nerve myelin. Here the regularly repeating unit is derived from the layers of membranes that are wrapped round the axon. The X-ray picture is composed simply of lines (Fig. 8.5). Because of its simplicity, both visually and mathematically, nerve myelin will be the only system analysed in detail here.

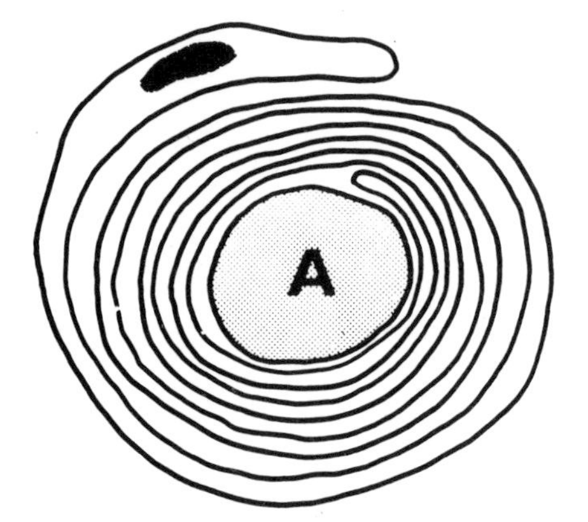

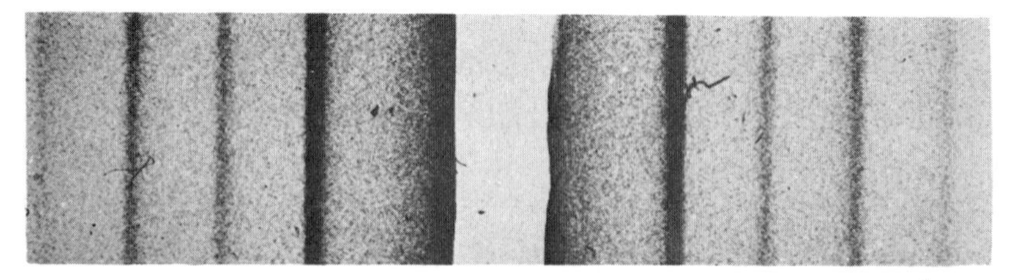

Fig. 8.5. Arrangement of Schwann cell membranes round the axon and the resulting diffraction pattern. (SFD = 1.16 m)

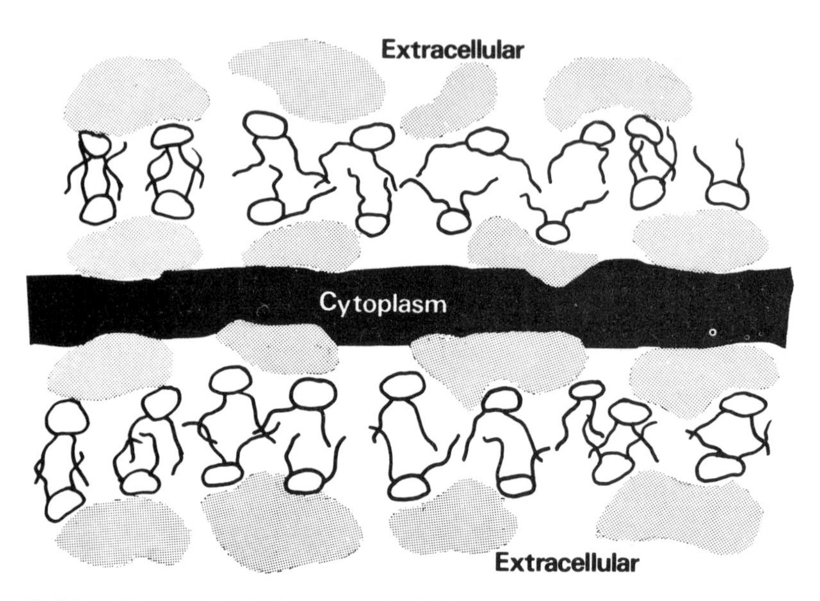

Fig 8.6a. Diagram of a single repeat unit of Schwann cell membranes. Hatched areas represent membrane proteins. Stain is taken up best by residual cytoplasmic material.

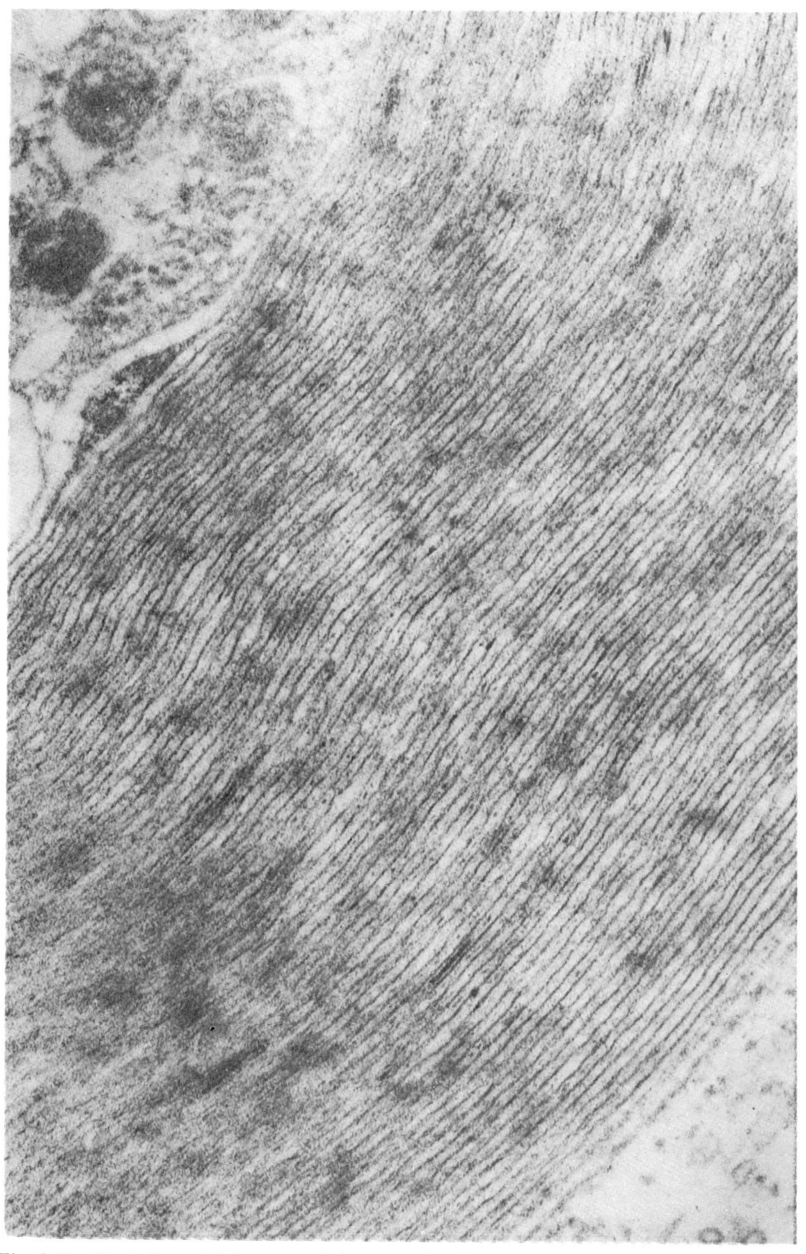

Fig. 8.6b Part of amphibian nerve Schwann cell. The cytoplasmic regions can be seen as very dark lines and the extracellular regions, which take up less stain, are the faint lines in between. From the given magnification, the membrane repeat distance can be calculated and compared with the X-ray values from Fig. 8.12. (Magnification 160,000 x)

Nerve myelin is formed from successive layers of Schwann cell membranes tightly packed together. A true repeating unit (as far as X-ray diffraction is concerned) consists of *two* membrane units together with a small amount of residual matter from former cytoplasmic and extracellular regions. The repeating unit can be seen in the electron microscope after staining with osmium tetroxide which is selectively taken up by the residual cytoplasmic proteins (Fig. 8.6b). These cytoplasmic regions are seen as dark lines in the electron micrograph. Sometimes visible between the dark lines are fainter lines representing the extracellular regions. We therefore have a quasi-crystalline structure in nerve myelin made up of repeating units of two membranes (Fig. 8.6a).

(ii) Description of a simple X-ray machine (low-angle)

All X-ray machines have three basic parts in common: (*a*) a *generator* to produce the X-rays; (*b*) a *monochromator* to provide radiation within a narrow wavelength band; and (*c*) a *'camera'* to record the diffraction patterns.

(*a*) In an X-ray generator, electrons are accelerated by high electric fields and strike a metal target (often copper) with high velocities. This produces a continuous spectrum with lines superimposed (Fig. 8.7). The accelerated electrons have sufficient energy to remove the copper electrons from their innermost orbitals (Fig. 8.2).

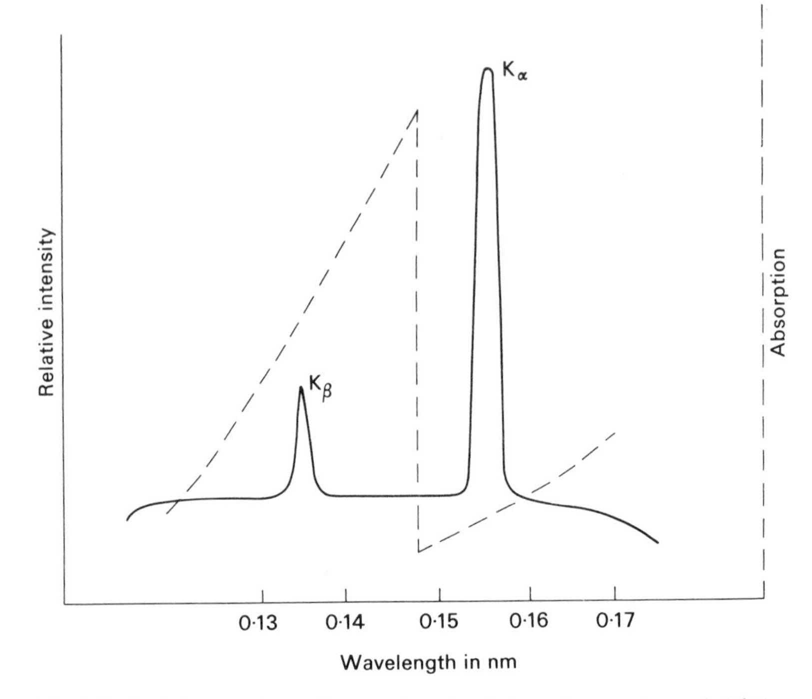

Fig. 8.7. Emission spectrum of copper target and absorption spectrum of nickel.

When a K-shell electron is lost there is the possibility of an L-shell electron jumping down to take its place and such a transition gives rise to the so-called K_α line of copper (wavelength 0.154 nm). The K_β line is involved with the $M \rightarrow K$ transition and as the energy step is larger the wavelength involved (0.139 nm) is smaller.

Monochromatic X-rays are required for clearly defined diffraction patterns and in older machines this was crudely accomplished by removing the K_β line with a thin metallic filter. Nickel, for example, has an absorption edge at 0.148 nm (Fig. 8.7) so it will absorb the K_β line and allow the K_α through. To the left of the absorption edge the photons have a high enough energy to eject electrons from the nickel K-shell, whereas to the right they do not have sufficient energy and the absorbance falls off. In modern machines filtering is achieved by means of a crystal monochromator rather than a nickel filter.

(*b*) The monochromator consists of a thin crystal which has been deformed to a curved shape so that the angle θ which the crystal planes make with the incoming rays is constant throughout the crystal (Figs. 8.8 and 8.9).

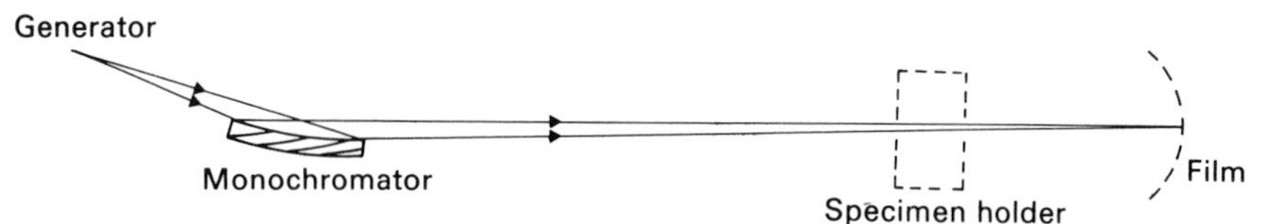

Fig. 8.8. Ray diagram showing path of X-rays from generator to film.

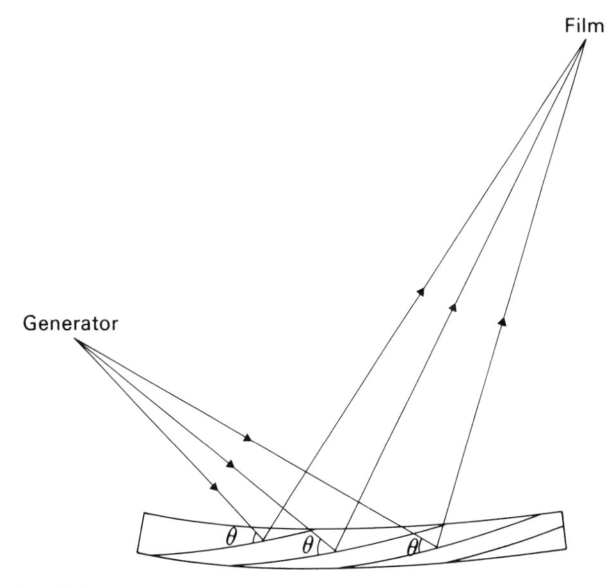

Fig. 8.9. Diagrammatic view of the crystal monochromator (the angle θ has been greatly increased for the sake of clarity). Only X-rays falling at an angle θ to the crystal planes reach the film.

A straight forward application of Bragg's equation shows why the X-rays are filtered

$$n\lambda = 2\, d \sin \theta \qquad 8.3$$

θ is set by fixing the crystal in position, and the interplanar spacing d is fixed by the nature of the crystal used so that X-rays of only one wavelength will pass through the system. By choosing the correct value of θ for the orientation of the crystal in the machine, only the K_α line will pass. The monochromatic X-rays are focused on the film because of the curved nature of the crystal.

(*c*) After leaving the monochromator the X-rays enter the camera which contains both the specimen to diffract the radiation and the film to record the pattern (Fig. 8.10). In more sophisticated machines the distribution of the diffracted energy is mapped out

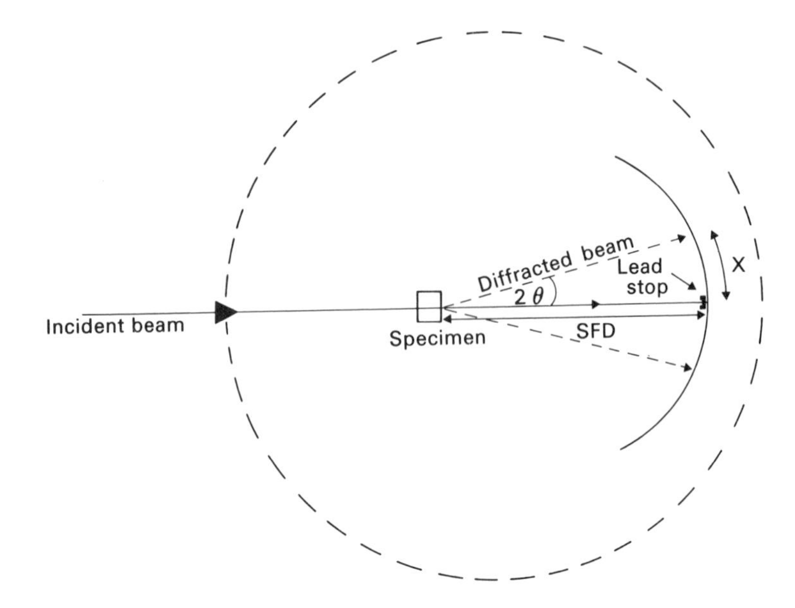

Fig. 8.10. X-rays enter the camera (a light-proof box shown in outline by dashes), are reflected from the crystal planes of the specimen and the diffraction pattern is recorded on film. A lead stop placed in front of the film prevents the very intense energy in zeroth order diffraction beam from fogging a large area of film. SFD is the specimen-to-film distance and X is the distance of the first order line from the straight through position. Note that 2θ is the angle between the undiffracted and diffracted beams. The scattering angles involved are small, i.e. low angle.

using radiation detectors (see Chapter 10, p. 176) such as Geiger-Muller tubes or scintillation counters.

(iii) Analysis of Diffraction Pattern

Bragg's equation must again be invoked to understand the pattern. In this case λ is fixed and d is fixed so θ is prescribed from these. Hence only those planes of membranes that are oriented at the angle θ to the incoming beam give rise to a diffraction pattern

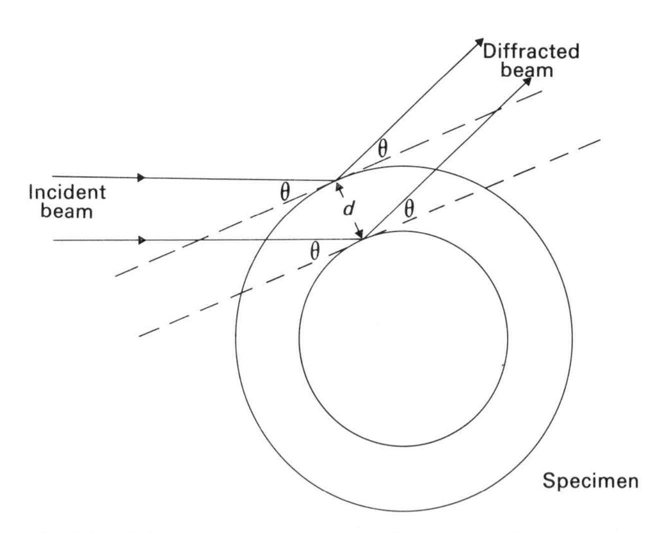

Fig. 8.11. Diagrammatic view of the diffraction of X-rays at nerve myelin. The X-rays are reflected at an angle θ to the membrane planes (parallel to the dashed lines) and interfere to form a diffraction pattern on the film.

(Fig. 8.11). When the X-rays diffracted at these planes constructively interfere, lines are observed corresponding to the first, second, third etc. diffraction orders.

A photograph of a diffraction pattern from nerve myelin is shown in Fig. 8.12 and from this you can calculate the repeat distance d.

$$d = \frac{n\lambda}{2\sin\theta} \tag{8.3}$$

and as the angles involved are small (low angle scattering)

$$d \approx \frac{n\lambda}{2\theta}$$

but $$2\theta = \frac{X}{SFD} \quad \text{(Fig. 8.10)}$$

hence $$d \approx \frac{n\lambda\,SFD}{X} \tag{8.4}$$

where SFD is the specimen to film distance and when $n = 1$, X is the distance from the straight-through position to the first-order line.

In a simple one-dimensional repeating structure, e.g. myelin, the values for successive values of X should be in the ratio 1:2:3:4 etc. In the fresh myelin pattern (Fig. 8.12a) you should find that they are in the ratio of 2:3:4 indicating that the first diffraction line is hidden behind the stop. When you are calculating the various values of d, therefore, you have to take $n = 2$ for the first measurable line, and $n = 3$ for the second and so on. In this way you should be able to obtain a mean value for the interplanar spacing d giving rise to several diffraction orders. The mean value derived from the several diffraction orders can then be compared with the d value measured from the electron micrograph of known magnification. As you may find that the two values are quite different, a further pattern from fixed nerve is provided (Fig. 8.12b) and you will be able to make some conclusions about the effects of fixatives on biological tissues.

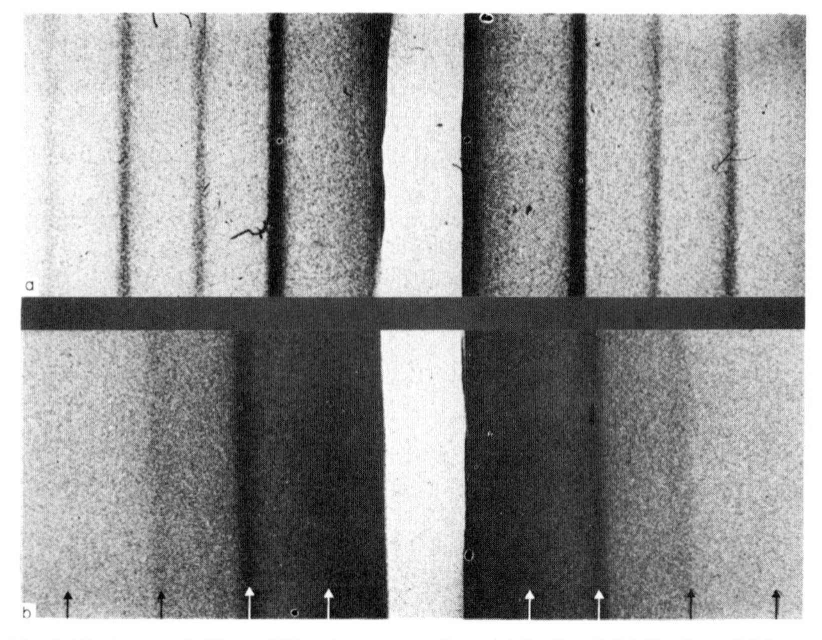

Fig. 8.12. Low angle X-ray diffraction pattern from (a) fresh and (b) fixed amphibian nerve (SFD 1.16 m). From the given specimen to film distance, the repeat distance for both patterns can be calculated and compared with the value from the electron microscope. In the case of fixed nerve the lines are much fainter (because there is some disruption of structure) and their positions are marked by arrows.

Careful examination of Fig. 8.12a will reveal that the lines from myelin have alternating intensities – why does this occur?
(Hint: What is the spacing of the diffraction lines from units that are spaced $d/2$ apart?)

8.4 Theory of Electron Microscope

The electron microscope is being increasingly used for high resolution work in the investigation of biological ultrastructure. It has very good resolution characteristics because of the small *wavelengths associated with highly energetic electrons*. This last statement might seem surprising at first, but it has been shown experimentally that diffraction patterns are produced when electrons strike an ordered structure in much the same way as light waves produce diffraction patterns. Hence, not only do light waves have some of the characteristics of particulate matter (photoelectric effect) but now we see that particles have wave characteristics.

The wavelength associated with an electron of energy E is given by

$$\lambda = \frac{hc}{E}$$

and in modern electron microscopes, as the electrons are accelerated by about 30,000 volts, the energy of the electrons will be 30,000 electron volts or $3 \times 10^4 \times 1.6 \times 10^{-19}$ J. Now, as c is 3×10^8 m s^{-1}, λ will be approximately 0.04 nm, and this is the theoretical limit of resolution of such an instrument. In practice, however, the resolution is limited by the specimen thickness and engineering difficulties in the design of the electrostatic lenses used to focus the electrons (Fig. 8.13). The best values obtained so far are in the region of 1 nm. Contrast is achieved through use of electron-dense stains, such as osmium tetroxide. These attach themselves to proteins in the cell and the

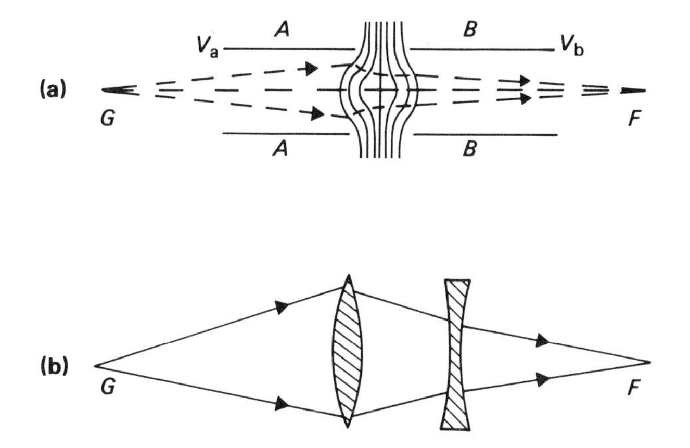

Fig. 8.13. (a) Electrostatic lens for electron microscope. The two cylinders A and B are at different potentials (Chapter 9) V_a and V_b where $V_b > V_a$. The electric fields between the plates, shown by the curved lines, will converge at F a beam of electrons that is diverging from G. (b) The optical analogue of the electrostatic lens shown in (a).

regions with the heavy metal osmium scatter electrons and so appear dark in contrast to the unstained regions which allow the electrons to pass (Fig. 8.6b).

8.5 Absorption of Light

The photoelectric effect (Fig. 8.1) demonstrated the dual nature of electromagnetic radiations. Under some circumstances, e.g., in the phenomena of light diffraction and interference, the resulting patterns can most readily be explained in terms of the wave theory. In other circumstances, when light interacts with matter and is then absorbed, and sometimes re-emitted, then the only explanation is in terms of the quantum theory. As biologists we are concerned mainly in the interactions of light with polyatomic molecules rather than simply atoms. The quantum theory states that electrons in molecules are arranged in energy levels, just as they are in atoms, the only complication being that there are many possible energy levels for the electrons (Fig. 8.14).

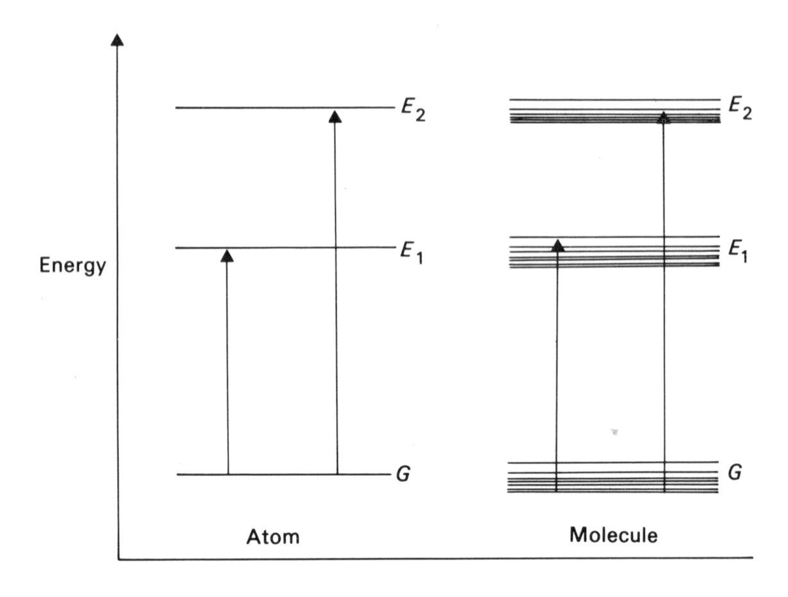

Fig. 8.14. Comparison of electronic energy states in atoms and molecules. G is the ground state and E_1 and E_2 are two possible excited states. In molecules the total number of energy states is increased because of molecular rotations and vibrations.

When light impinges on matter, the probability of it being absorbed depends upon the energy of the photon and the distribution of energy levels in the matter. The photon only has a significant probability of being absorbed if there are two energy levels whose energy difference exactly matches the energy of the photon.

In some substances, e.g. glass, the difference between energy levels is so great that the relatively low energy visible light photons are not absorbed and the material is transparent. Good conductors of electricity on the other hand have electrons that are free to move into a vast range of energy states so that almost every incident photon is likely to be absorbed. However, in most cases the photons are rapidly re-emitted in a backward direction and as a result thin metal sheets appear quite opaque as all the photons are reflected.

When light falls on a suspension of biological molecules, a certain fraction of the light is absorbed, and the amount absorbed depends both on the molecules comprising the suspension and the wavelength of the light. When the absorption versus wavelength characteristics are analysed with the aid of a *spectrophotometer* (section 8.6) several

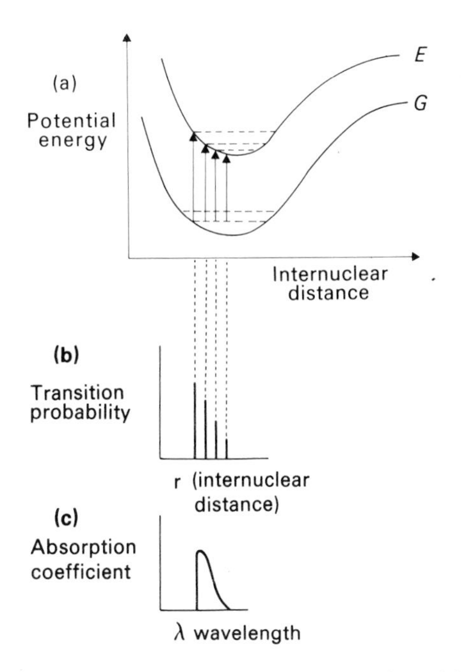

Fig. 8.15. (a) Diagrammatic representation of the vibrational energy levels in a diatomic molecule. G and E are the ground state and excited state respectively. (b) Five possible electronic transitions are shown and their transition probabilities are given. As nuclei spend most of their time at the extreme position of the vibration, the furthest left transition is the most likely. (c) An absorption band is finally observed because rotational energy levels broaden the individual lines. (After Thomas, 1965) Reprinted by permission of Associated Scientific Publishers.

absorption bands are often evident. The reason why there are bands rather than lines lies in the fact that polyatomic molecules can no longer be treated as small rigid particles.

The rotation of the molecule and the vibrations of its constituent nuclei must be taken into account. The energies of these motions are also quantized and resulting energy levels must be added to the atomic energy levels to obtain a complete picture of the available energy states.

A picture of how the vibration levels may be built up can be obtained most simply from the consideration of a diatomic molecule. In the lowest energy state the amplitude of vibration of the nuclei is small, and as energy is pumped into the molecule, e.g. by heating, then the amplitude of vibration will increase and other energy levels will be created. In this context the quantum theory states that only certain amplitudes of vibration are allowed and so discrete energy levels are created. A graph of potential energy against internuclear distance has a roughly parabolic form (curve G in Fig. 8.15a), and possible vibrational energy levels are shown for the electronic ground state of a diatomic molecule by the dashed horizontal lines. In the same manner a curve can be drawn for an electronically excited molecule E. When photons are absorbed, electrons will jump from a ground state energy level to an excited state level. Let us suppose for simplicity that only the lowest energy level is occupied in the ground state. The very important Franck-Condon principle states that as electrons can move faster than nuclei, the latter do not have time to move during a transition, and so the absorption act (which takes approximately 10^{-8} sec) can be represented by a vertical arrow. As nuclei, spend most of their time at the extreme position of their vibration, then the most probable transition is represented by the arrow furthest to the left (Fig. 8.15a). The other transitions are less likely. The probability of a transition from the ground state, i.e. the probability that a photon will be absorbed, therefore depends on the internuclear separation (Fig. 8.15b) and as this separation determines the energy of the molecule, the transition probability will also depend on the energy of the photon (and hence on its wavelength). Because the molecules have rotational as well as vibrational energy the individual lines are broadened to give an *absorption band* (Fig. 8.15c). Here we have dealt with only transitions from the lowest vibrational energy level of the ground state at room temperature, but at least the next level above this should also be considered. The complete absorption spectrum will therefore be more complex than that shown in Fig. 8.15c. It will in fact also have minor absorption bands on the short wavelength side of the absorption maximum.

Although the true state of affairs is much more complex in polyatomic molecules, absorption takes place along much the same lines. When a molecule has absorbed a photon it is in an excited state and has an excess of energy which it must be rid of, and there are several means of achieving this end. In the transitions $G \rightarrow E_2$ and $G \rightarrow E_3$ (Fig. 8.16) the electrons return within 10^{-12} seconds to the excited state E_1 and in these transitions energy is lost in the form of heat. The electrons remain in E_1 for over 10^{-8} seconds and when they eventually return to the ground state they emit light, in fact they *fluoresce.* This internal rearrangement of electrons can be seen in chlorophyll which, though it absorbs in the blue and in the red, only fluoresces in the red.

Electrons can become trapped in excited energy levels called triplet states, from which they cannot easily drop to the ground state. When the electron from an excited triplet state drops to the ground state, the light emitted is called *delayed fluorescence* if it takes place within 10^{-3} seconds, otherwise it is known as *phosphorescence*. Molecules which possess these triplet states are highly reactive and the excitation energy can be passed on to other molecules if they have matching energy levels. Chemical energy, e.g. in the form of dissociation energy, can be passed on in this way.

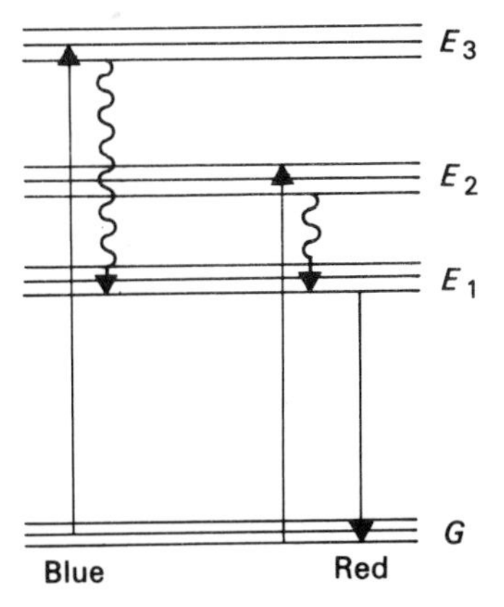

Fig. 8.16. The energy levels of a polyatomic molecule. Part of the energy absorbed in the transition $G \rightarrow E_3$ is lost due to radiationless transfer through overlapping levels to E_1. The energy content of the photon emitted in the transition $E_1 \rightarrow G$ is less than that absorbed and hence the wavelength is shifted from blue → red.

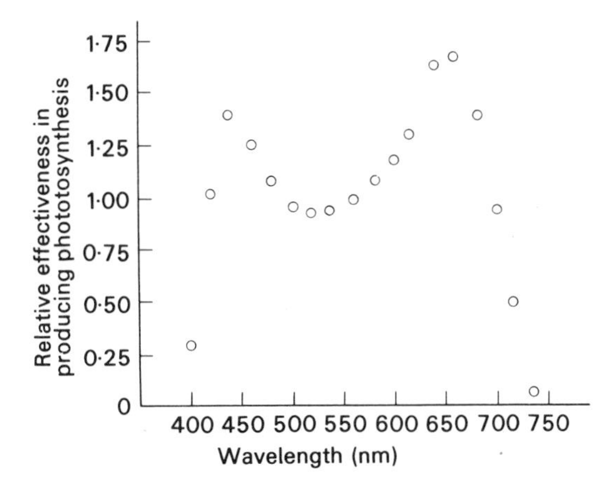

Fig. 8.17a. The effectiveness of various wavelengths of light in producing photosynthetic carbon dioxide fixation. Two clear peaks are in evidence. The fact that the effectiveness does not go down to zero between the peaks indicates the existence of at least one more peak in the wavelengths between the two. (Epstein, 1963)

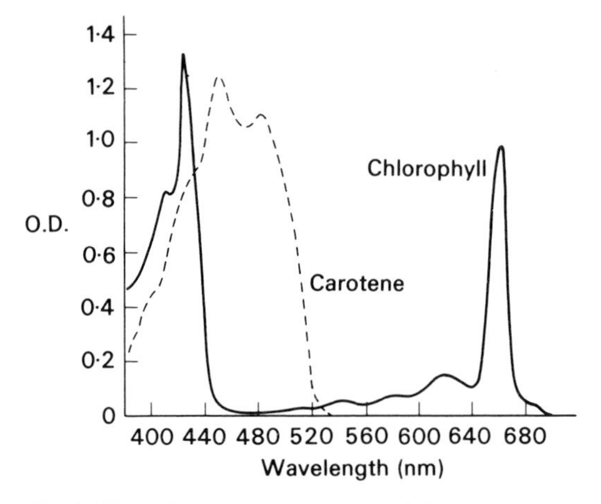

Fig. 8.17b. The absorption spectra of chlorophyll and carotene. The chlorophyll peaks at about 430 and 665 nm and the carotene peaks near 450 and 490 nm are identified as being responsible for the action spectrum peaks in the preceding Figure. (Epstein, 1963) Reproduced by permission of Addison–Wesley Inc.

The coupling between carotenoids and chlorophyll is an example of energy transfer from molecule to molecule. That the two pigment systems are involved in photosynthesis is shown by the fact that photosynthetic effectiveness, e.g. measured in terms of CO_2 fixation, does not fall to very low values in the region between two chlorophyll absorption peaks (Fig. 8.17a and b). It was therefore proposed that the energy absorbed by carotene can be passed on to the chlorophyll. This hypothesis is further backed up by spectroscopic evidence which shows that when organisms are irradiated with 500 nm light (absorbed only by the carotenoids) there is no carotenoid fluorescence, but instead there is the typical red fluorescence from chlorophyll.

8.6 Absorption Spectrophotometry

(i) The instrument

Newton's fundamental discovery (Chapter 7, p. 85) that white light on refraction is dispersed into its constituent colours is the basis of the prism spectrophotometer.

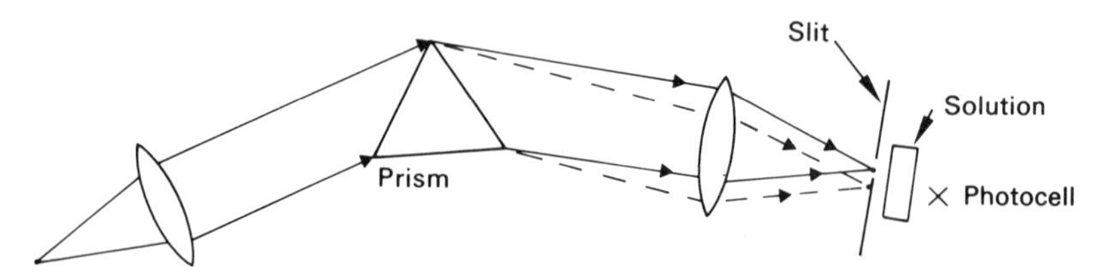

Fig. 8.18. Spectrophotometer, with prism monochromator. Blue light (dashed) is refracted more than red (solid) and so with the prism mounted on a rotating table, a narrow range of wavelengths can be selected by the slit.

The parallel beam of white light from a source will be dispersed into its constituent colours by the prism and these will be focused at different points on the focal plane of the lens (Fig. 8.18). The chosen wavelength can be made to fall on the slit by rotating the prism and after passing through the sample the intensity of the light is measured by means of a photocell. Spectrophotometers are required to work over a wide wavelength range and there are basically 4 technical problems to overcome.

(a) *Suitable light source.*
In order to overcome the first difficulty, two lamps have to be used to cover the necessary wavelength range. Usually they are a tungsten lamp producing a broad spectrum to cover the range 1000 to 320 nm and a hydrogen lamp, which produces narrow emission lines, is used from 320 to 190 nm.

(b) *Means of producing a pure spectrum.*
Either prisms or diffraction gratings (section 7.7) are used to obtain a pure spectrum. Light of the required wavelength is obtained by rotating the prism, or grating, and the rotating head carries a calibrated wavelength scale.

(c) *Material with which to make lenses and absorption cells.*
As glass absorbs strongly in the lower wavelength regions, the prism, lenses and test cells have to be made from silica.

(d) *Means of detecting light transmitted through specimen solution.*
Photocells are normally used to detect the light passing through the sample and the output from these is amplified and the result is displayed on a scale, or chart paper calibrated in terms of percent transmission or optical density.

(ii) Theoretical treatment

Absorption is a matter of probability, and in order that the statistical laws hold, we must assume this treatment will be confined to dilute solutions.

Suppose that we have a light beam of intensity I which after passing through the test cell of width ΔX, holding a solution, falls on a photocell which measures the transmitted energy $(I - \Delta I)$. If we carried out this experiment, then we would find that the amount of light absorbed ΔI was proportional to I and this is reasonable because the more photons there are in the incidental beam, the greater probability there is of one being absorbed. Provided the width ΔX was small, then ΔI would also be proportional to ΔX. ΔI would also depend on the substance in the solution, e.g. on whether the material had an absorption band at the wavelength used, and the constant for the material μ is called the *absorption coefficient.*

Hence we can write

$$\Delta I = -\mu I \Delta X$$

Now if we allow ΔX to become infinitely thin

$$dI = -\mu I\, dx$$

or

$$dI/dx = -\mu I$$

and integrating this equation from $x = 0$ to $x = x$ gives

$$I_x = I_0 \exp(-\mu x) \qquad 8.5$$

where I_0 is the intensity incident on the cell, and I_x is the intensity at some distance x from the surface.

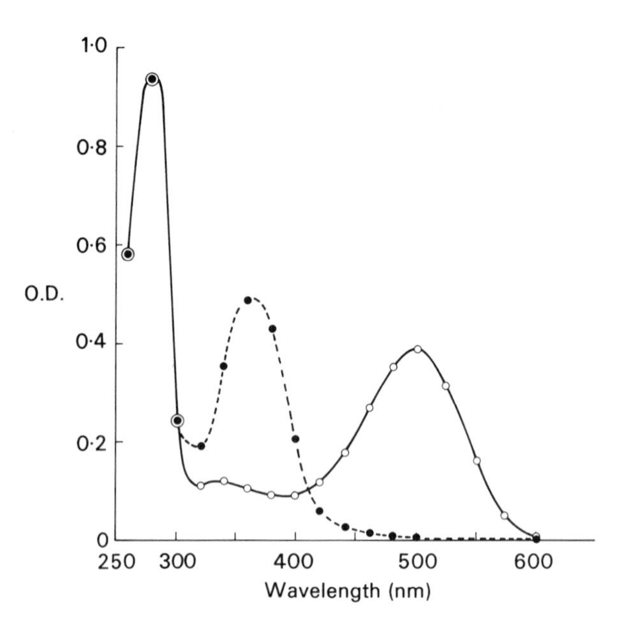

Fig. 8.19. Absorption spectra of cattle rhodopsin in 1% digitonin, before (solid) and after (dashed) illumination in the presence of hydroxylamine (Data kindly provided by Professor S. L. Bonting).

As absorption is a probability phenomenon, then the more absorbers there are in the light path, the more will be absorbed. For dilute solutions, the absorption coefficient will be proportional to the concentration, C, i.e.

$$\mu = \beta C$$

and β is known as the *extinction coefficient.* Sometimes the *transmittance* is measured, and this is simply the ratio I_x/I_0. Because of the very wide range of transmittances possible, e.g. from 0.1 to 100%, it is usually more convenient to display the result on a logarithmic scale and in this way the *optical density* D is measured. D is defined by the equation

$$D = \log_{10} \frac{I_0}{I_x}$$

and as $$\mu = \frac{1}{x} \log_e \frac{I_0}{I_x} \qquad 8.5$$

then $$D = \mu x / 2.303 \qquad 8.6$$

and $$C = 2.303\, D/\beta x \qquad 8.7$$

In absorption spectrophotometry, a further quantity is often used, the *molar extinction coefficient* ϵ_{mol}, defined by the relationship

$$\epsilon_{mol} = \frac{D}{Cx} \qquad 8.8$$

The units invariably used are not SI units but because their use is so widespread they will be given here: C has units mol litre^{-1} and the cell path length x is in centimetres, then ϵ_{mol} has the units litre mol^{-1} cm^{-1}.

Problem 8.1

Photoreceptor cells (Rod Outer Segments, Fig. 10.7) were isolated from cattle retinae, solubilized in digitonin and the absorbance (optical density) read before and after bleaching (Fig. 8.19). Given that the total dry weight concentration of the receptors was 2.7 mg/ml, that the molar extinction coefficient of rhodopsin at 500 nm is 41,000 litre mol^{-1} cm^{-1}, that the molecular weight of rhodopsin is 40,000 and that a 1 cm spectrophotometer cell was used, show that rhodopsin represents a significant fraction of the total outer segment dry weight.

References

Ackerman E. (1962) *Biophysical Science.* Prentice-Hall, New York.
Clayton R.K. (1965) *Molecular Physics in Photosynthesis.* Blaisdell, New York.
Epstein H.T. (1963) *Elementary Biophysics.* Addison-Wesley, Reading, Mass.
Harrison G.R., Lord R.C. & Loofbourow J.R. (1960) *Practical Spectroscopy.* Prentice-Hall, New York
McKenzie A.E.E. (1959) *A Second Course of Light.* Cambridge University Press.
Setlow R.B. & Pollard E.C. (1962) *Molecular Biophysics.* Pergamon, London.
Thomas J.B. (1965) *Primary Photoprocesses in Biology.* North Holland, Amsterdam.
Wilson H.R. (1966) *Diffraction of X-rays by Proteins, Nucleic Acids and Viruses.* Arnold, London.

Chapter 9
Electricity

9.1 Introduction

At present, the two most important areas of interest for a biologist studying electricity are *electrostatics* and *current electricity*. The former is required to understand the interaction of charged species, ions and molecules, at the molecular and cellular level. The latter is essential in order not only to understand the electrical signalling processess that take place along both animal and plant membranes, but also to use intelligently the vast range of electronic equipment available to biologists of all interests. The practical basis of the design and use of such equipment will not be dealt with here as it forms a large and specialized field and those specially interested are recommended to read Donaldson (1958) and Young (1973). For much the same reason, *electromagnetism* is only given a cursory treatment.

9.2 Basic Concepts

The study of electricity is concerned with the behaviour of charged particles, electric fields and magnetic fields. It is unfortunate that there is no simple intuitive model for the easier understanding of these terms. An *electric field* is defined in terms of its action on a charge and an *electric charge* is defined in terms of its behaviour in an electric field. The tautological nature of these concepts is extended to the definition of a magnetic field, which can neither be produced nor detected without the aid of moving charges. Charge has a positive or negative sign. Two bodies, A and B, with charges of the same sign exert a repulsive force on one another (Fig. 9.1) and consequently work must be done to bring them together. The converse holds with bodies of opposite sign.

9.3 Coulomb's Law

The experimental relationship found for charged spheres (Coulomb's Law) is

$$F = \frac{q_1 q_2}{4\pi K \epsilon_0 x^2} \qquad 9.1$$

where F is the magnitude of the force between the spheres, x is the distance between the centres of the charged bodies. The units of charge are coulombs (C) and charge is an integral multiple of the charge on a single electron ($e = 1.6 \times 10^{-19}$ C). When q_1 and

q_2 have the same sign the force exerted is a repulsive one (Fig. 9.1). ϵ_0 is a coefficient called the *absolute permittivity* of free space and has the value $8.85 \times 10^{-12}\ C^2\ m^{-2}\ N^{-1}$. K is the *relative dielectric constant* of the medium separating the charges and has the value of 1 for a vacuum and 80 for water, for example (section 9.22).

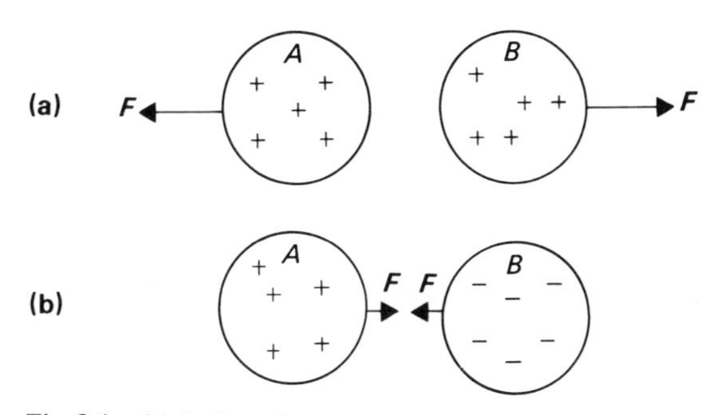

Fig. 9.1. (a) Bodies of similar charge repel one another. (b) Bodies carrying opposite charges attract. The force between bodies is given by Coulomb's Law (section 9.1).

Like gravitational attraction this force is of the action-at-a-distance type, making itself felt without the presence of a material connection between A and B. No one knows why this is possible, it is simply an experimental fact. It is useful to think of each of the charged bodies as modifying the space around it, and setting up an electric field.

9.4 Electric Field

In Fig. 9.1a, if body B is removed, then A is still said to exert an *electric field* at the point P (Fig. 9.2). To show that an electric field exists at P, one simply places a small test charge (q) there and if a force is exerted, then a field is said to exist at P.

Fig. 9.2. The fact that a force is exerted on a small test charge placed at P indicates the presence of an electric field there. The field vector is in the same direction as the force vector when the test charge is positive.

The field vector **E** is defined by the relation

$$\mathbf{E} = \frac{\mathbf{F}}{q} \qquad 9.2$$

when $F = 1$ N and $q = 1$ C, then $E = 1$ N/C.

Suppose a small charge q moves from a to b under the influence of a field $\mathbf{E}$, then the work done W in moving this distance is given by

$$W = \int_a^b \mathbf{F} \,.\, ds = q \int_a^b \mathbf{E} \,.\, ds \qquad 9.3$$

Note that the work done is the integral of a vector product, i.e.

$$W = q \int_a^b E \cos \theta \; ds \qquad \text{(Appendix II)}$$

It can be shown that the work done in an electric field is conserved, i.e. $\int_a^b E.ds$ is independent of the path taken. When a force has this property (for example, gravitational forces also have this property) then the work done on the particle is the difference in potential energy of the particle between the end point and the starting point, i.e.

$$W = U_b - U_a$$

i.e.

$$U_b - U_a = q \int_a^b \mathbf{E} \,.\, ds \qquad 9.4$$

where U_a and U_b are the initial and final potential energies respectively.

9.5 Electric Potential

Instead of dealing directly with the potential energy of a charged particle it is useful to introduce the more general concept of energy per unit charge. This quantity is called *potential, V*. Thus for a particle of charge q and potential energy U_p,

$$V = \frac{U_p}{q} \qquad 9.5$$

and thus from equation 9.4,

$$V_b - V_a = \int_a^b \mathbf{E} \,.\, ds \qquad 9.6$$

$V_b - V_a$ is the potential difference between the two points a and b. It has the units of $\mathrm{J\,C^{-1}}$. One $\mathrm{J\,C^{-1}}$ has the special name of one *volt.* A value can be assigned to the potential at a single point only when some arbitrary reference point is selected at which the potential is zero. In the case of the potential difference across a cell membrane for example, it is usually convenient to regard the outside medium as the reference point.

9.6 Summary

The interaction of two systems of charges can be characterized in several ways (Fig. 9.3).

(i) There is a force vector $\mathbf{F}$ on a charge q in the vicinity of a system of charges S.

(ii) There is a vector field **E** at a due to the presence of the system S.
(iii) Work is done when q moves from a to b and the work is independent of the path taken.
(iv) There is an electrical potential difference between the points a and b, given by the work done in moving a unit positive charge from a to b.

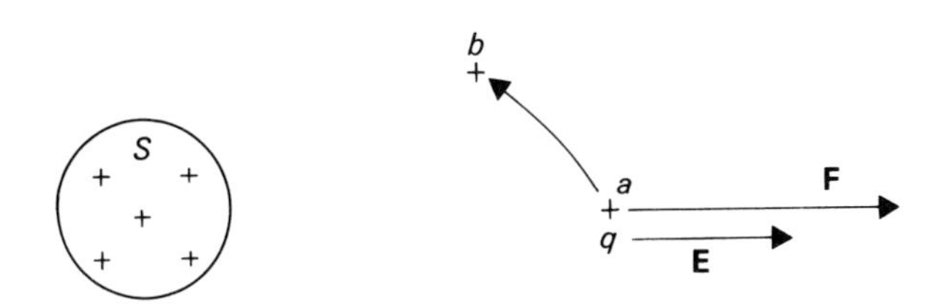

Fig. 9.3. The interaction of two systems of charges can be described in several ways. (See text for explanation)

9.7 The Cathode Ray Oscilloscope

The Cathode Ray Oscilloscope is widely used in electrophysiology when rapidly changing electrical signals are studied, e.g. nerve action potentials. The mode of operation of the cathode ray tube itself (Fig. 9.4) provides a simple example of the movement of charges, electrons, under the influence of electric fields.

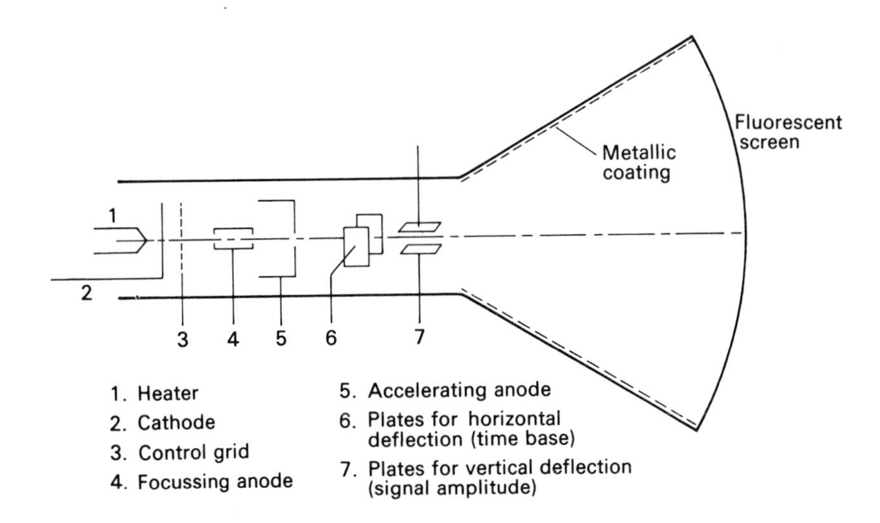

Fig. 9.4. The cathode ray tube. Electrons are accelerated towards the screen by an electric field between the cathode (2) and anode (5).

The whole interior of the tube is highly evacuated. The cathode at the left is raised to a high temperature by the heater and electrons, cathode rays, evaporate from its surface. The *accelerating anode* is maintained at a high positive potential V, relative to the cathode so that there is an electric field directed from right to left between the anode and cathode. This field is confined to the cathode-anode region and electrons passing through the hole in the anode travel with a constant velocity from the anode to the fluorescent screen. The function of the control grid is to regulate the number of electrons that reach the anode and hence the brightness of the spot on the screen.

The complete cathode-anode assembly is called the electron gun. The electrons then pass between two pairs of deflecting plates, the first of which controls the horizontal deflection of the beam (Fig. 9.5a). The sweep speed is controlled by the potential difference across the plates which varies in time with a sawtooth waveform (Fig. 9.5b). The electrons arrive

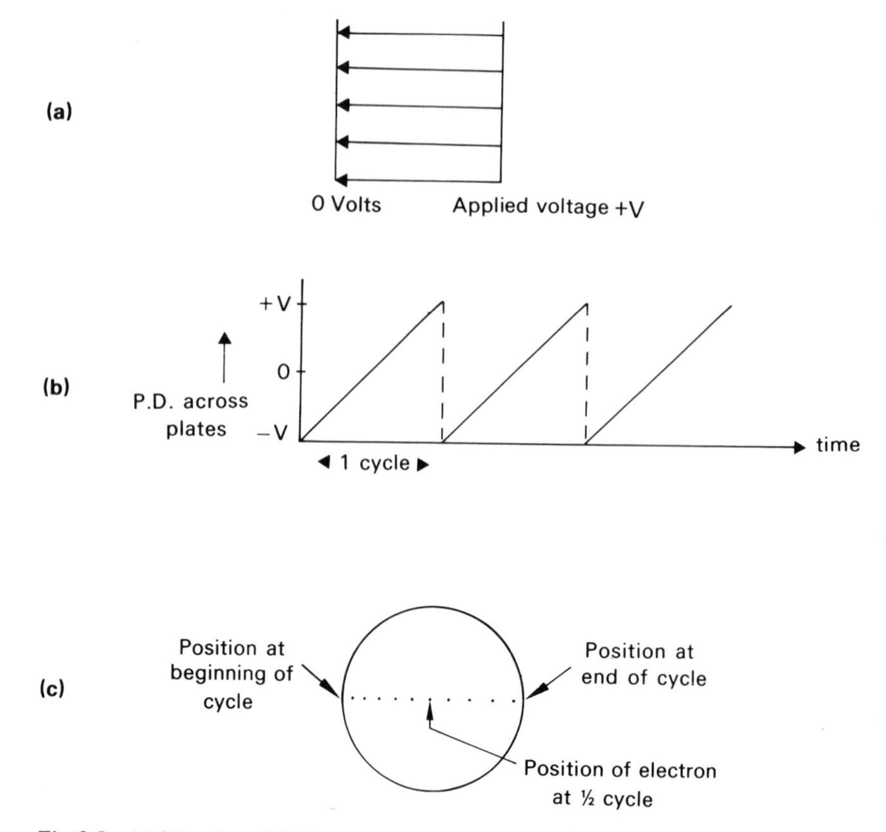

Fig. 9.5. (a) Direction of field vectors between two parallel metal plates with a potential difference of $+V$ volts between them. In this case $E = V/d$ where d is the distance between the plates. (b) Variation of p.d. across the horizontally deflecting plates with time. (c) Showing how one sweep is built up from the successive arrival of electrons at the fluorescent screen.

initially at the left hand side of the screen and, as the field decreases, they are deviated less and so they arrive at the centre of the screen when the field across the plates is zero halfway through the cycle. Then the field increases in the opposite sense. At the end of the cycle, the field across the plates reverts to its initital value. The field across the vertical plates is controlled by the potential difference across the source under investigation and hence a two dimensional image is formed on the screen. This can be then photographed for detailed study later (Fig. 9.6).

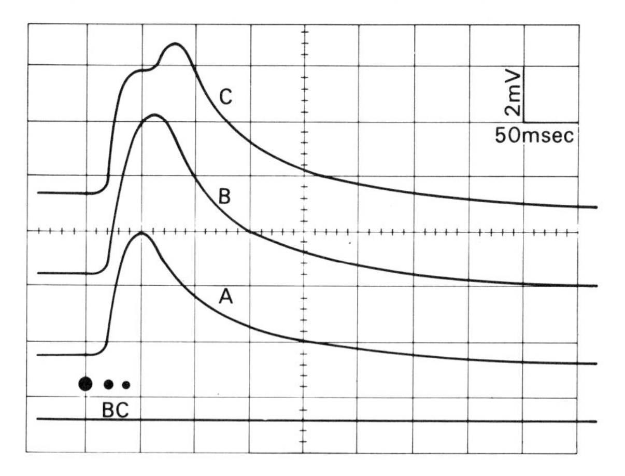

Fig. 9.6. Example of a photograph of an oscilloscope trace, showing the experimental response of the potential difference across *Sepiola* retina to 10 μs light flashes. Bottom trace: stimulus markers. The first marker is the same for each of the traces above. A = response to single flash; B = response to two flashes 20 ms apart; C = response to two flashes 40 ms apart. (From Duncan & Croghan, 1973)

9.8 Electric Fields and Sense Organs

Certain types of electric fishes, of which *Gymnarchus* is a striking example, are able to locate objects by sensing changes in the electric fields set up by the fish themselves (Fig. 9.7).

The fish have stacks of modified tissue called electric organs set up in regular array along their bodies and through these maintain a considerable electrical potential difference between the head and tail. The field is pulsed by the fish and 25 ms is the average length of a pulse of which there are about 7 or 8 per second. The skin of the fish probably acts as an insulator except in certain regions near the head where jelly-filled pits provide conducting paths for the electric field. These pits have some type of sense organ at their base and it is these receptors that can detect very small changes in the field distribution at the head. With its detectors a fish can detect the presence of a conducting rod in its tank even when it is only 2 mm in diameter. These fish are able to live in muddy streams where the use of eyes for catching prey or avoiding predators is impossible.

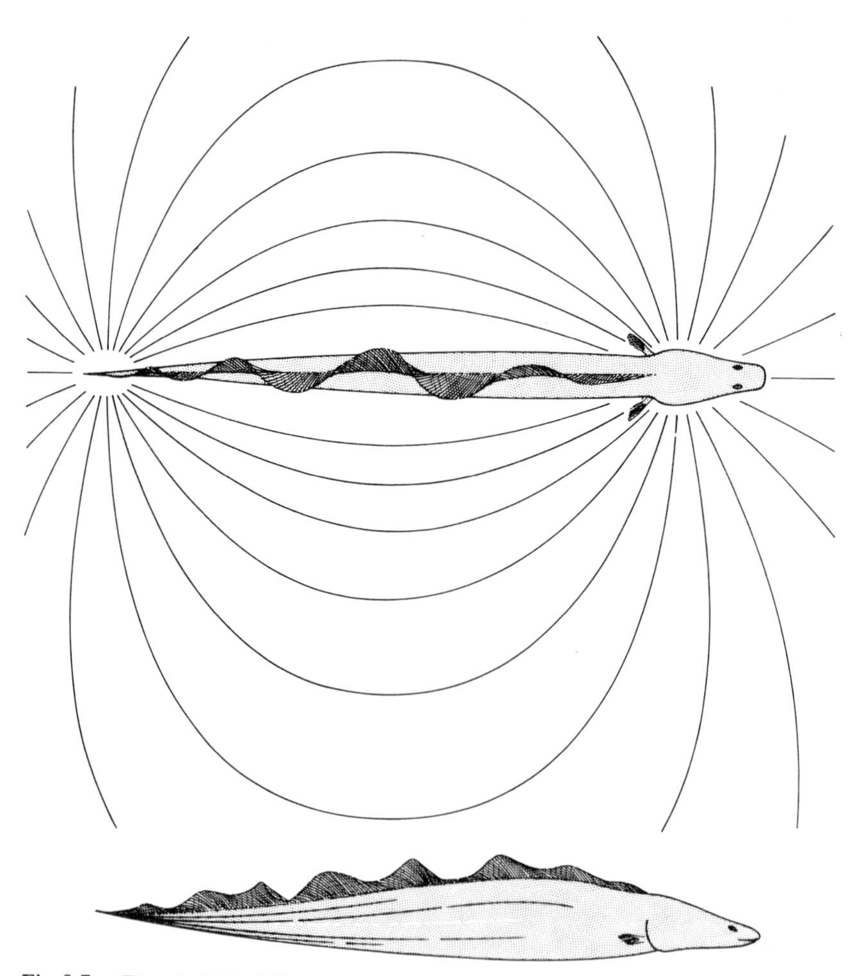

Fig. 9.7a. Electric field of *Gymnarchus* and location of electric generating organs are diagrammed. Each electric discharge from organs in rear portion of body makes tail negative with respect to head. Most of the electric sensory pores or organs are in head region. The undisturbed electric field resembles a dipole field, as shown, but is more complex. The fish responds to changes in the distribution of electric potential over the surface of its body. The conductivity of objects affects distribution of potential. (From Lissman, 1963)

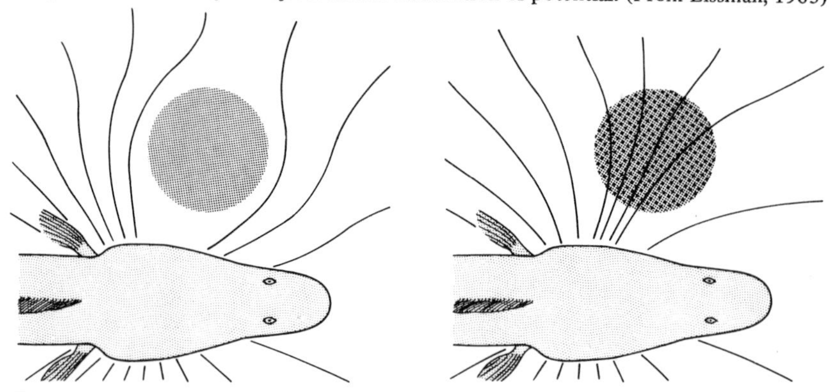

Fig. 9.7b. Objects in electric field of *Gymnarchus* distort the lines of current flow. The lines diverge from a poor conductor (left) and converge toward a good conductor (right). Sensory pores in the head region detect the effect and inform the fish about the object. (From Lissman, 1963)
Copyright © (1963) by Scientific American Inc. All rights reserved.

9.9 Electromotive Force

A body with a considerable excess of electrons has a strong negative charge and one with a deficit has a positive charge. Between them there exists a field as long as they are separated. If some material allowing free movement of electrons is placed between two bodies carrying different charges, electrons move through this conductive path. If the movement of electrons is to be maintained, then a continuous surplus must be generated to replace electrons moving away through the conductor. The surplus can be maintained chemically, as in a battery, or electromagnetically as in a generator. These systems are termed sources of *electromotive force* (EMF).

One common example of a battery is the *primary cell* (Fig. 9.8) which consists of zinc and carbon plates immersed in an electrolyte solution of ammonium chloride. Zinc

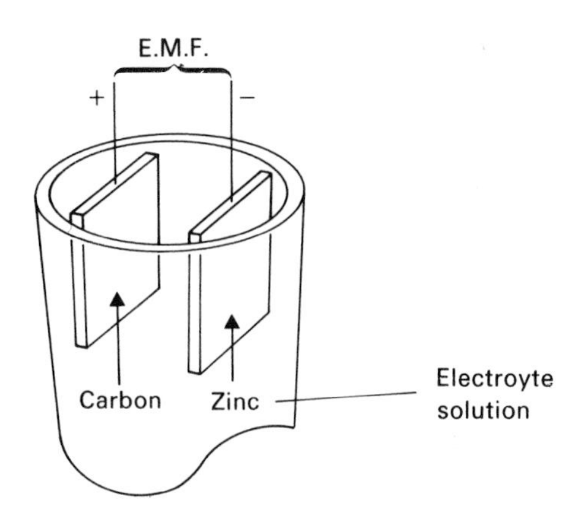

Fig. 9.8. Primary cell.

goes into solution as the divalent ion Zn^{2+}, leaving two surplus electrons behind so that a negative charge is left on the zinc plate. The solution becomes positively charged and so does the carbon plate in contact. When a wire connects the two terminals, *current* flows and this is maintained by a continual dissolving away of the zinc plate. The EMF developed in this type of cell is about 1.5 volts. When drawing circuit diagrams, a source of EMF is represented by ||, the long line denoting the positive terminal (carbon) and the shorter line the negative (zinc) terminal.

If batteries are connected in *parallel* (Fig. 9.9a) there is more area for chemical action and a more powerful current is available but the potential available remains the same. However, when they are connected in *series* (9.9b), the voltage steps are additive and a higher overall voltage is available.

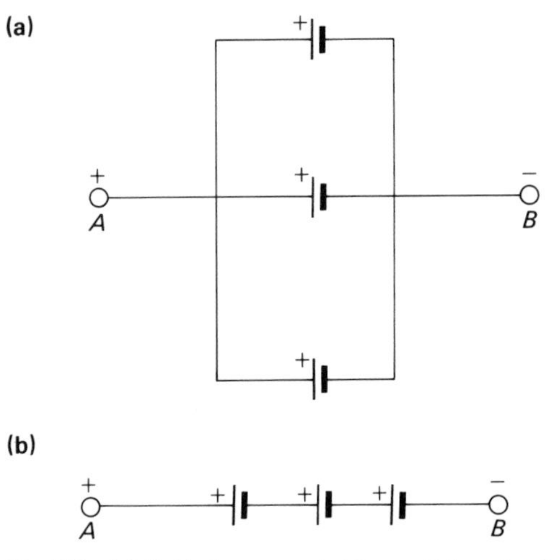

Fig. 9.9. (a) Batteries connected in parallel, each of EMF 1.5 volts, then the total open circuit EMF ($V_A - V_B$) is still 1.5 volts, but more current will be available. (b) When the batteries are connected in series, the open circuit EMF is now 4.5 volts, but the total current available will be the same as for one alone.

9.10 Electrical Current

Electrical current, as mentioned above, is the flow of charge and is defined by the relationship

$$I = \frac{dq}{dt} \qquad 9.7$$

where the current I has the units Cs^{-1} which are given the special name *amperes*. The current is the rate at which coulombs are passing a given point in a system. As charge has a sign, current will have a sign. It is conventional to regard the direction of flow of current as the direction of flow of positive charge, whatever the sign of the charge of the moving particles. In the case of current flow across a cell membrane, a further arbitrary definition of sign must be made. The convention that will be used here is that a positive membrane current is a current flowing into the cell.

9.11 Electrical Conductance

Certain substances allow currents to pass relatively easily and these are termed *conductors*. Other substances offer much more resistance to the flow of current and these are termed *insulators*. A part of a circuit whose resistance is negligible is called a *short circuit*; one whose resistance is effectively infinite is called an *open circuit.*

In the case of a metallic solid, the outer, valency, electrons are loosely held to the atomic nuclei and can move freely about the crystal lattice as a sort of electron gas. Such a substance is a good conductor. If such a substance is heated up, electrons can be driven from the surface. Heated cathodes are used to provide free electrons in valves cathode ray tubes, X-ray tubes and electron microscopes.

Another type of conduction is *ionic conduction*. Here the mobile charged particles are ions which are much larger particles than electrons and which have either a negative, *anion*, or positive, *cation*, charge. These can only move significantly if the medium is of relatively low viscosity. This is the mechanism of conduction in electrolyte solutions and probably in biological membranes.

Insulators are substances with few mobile electrons or ions. However, there is no sharp distinction from conductors. Crystalline solids showing relatively poor electronic conduction are termed *semiconductors*. Examples are germanium, silicon, selenium. These have a number of important practical applications, e.g. thermistors, photoelectric cells, rectifiers, transistors.

Insulators are often exposed to enormous electric fields and the materials of which they are composed are subject to enormous stress. As the field is increased there comes a point when there is a physical breakdown of the insulator, *dielectric breakdown*, and a surge of current flows through the insulator. The maximum field an insulator can support is termed its *dielectric strength*. The potential difference across a cell membrane is usually about 0.1 V, inside negative and as membranes are only of the order of 5 nm thick, the field across the membrane is thus about 2×10^7 Vm^{-1}. Artificial membranes undergo dielectric breakdown at a potential difference of about 0.25 V, suggesting that the cell membrane is operating at a point relatively near breakdown.

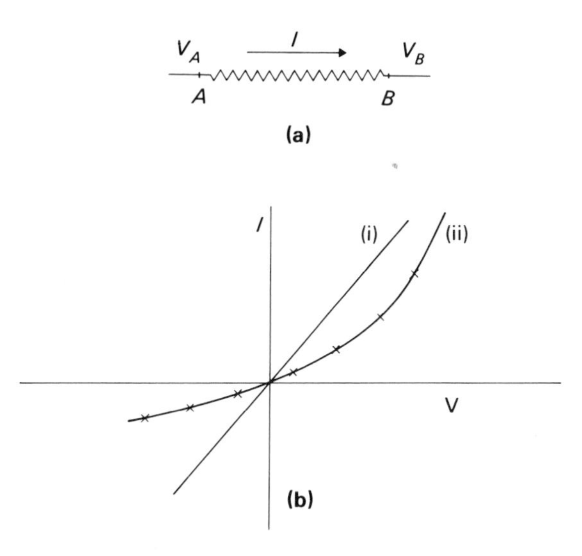

Fig. 9.10. (a) *AB* represents a resistor. V_A and V_B are the potentials at A and B respectively when a current I flows. (b) Current-voltage curves for (i) an ohmic resistor, e.g. a copper wire and (ii) a non-ohmic resistor, e.g. squid nerve membrane.

9.12 Ohm's Law

AB represents a conductor (Fig. 9.10a) and the experimental relationship between the current through the conductor and the potential drop across it was discovered by Robert Ohm.

$$I = \frac{1}{R}(V_A - V_B) \qquad 9.8$$

or $$I = G(V_A - V_B) \qquad 9.9$$

where R is the resistance across AB and has units $V\,A^{-1}$, also given the special name *ohm* (shorthand symbol: Ω); G is the conductance and has units Ω^{-1} (also mho). If the current voltage curve of a conductor is linear (Fig. 9.10b), then the behaviour is described as *ohmic*. The curves obtained from many biological membranes, e.g. from nerve membranes are non-ohmic, and are called *rectifying*. In this case the resistance to current passing in one direction is greater than the resistance to current flowing in the opposite direction. The conductance, or resistance, is in fact, a function of potential.

9.13 Temperature and Resistance

The resistance of an ordinary metallic conductor increases with increasing temperature. Thus a resistor placed in a resistance measuring circuit can, after calibration, be used to measure temperature. Semiconductors show a particularly large temperature effect with the resistance actually decreasing with increasing temperature. Such a semiconductor device is known as a *thermistor* and because of its small size and thus rapid response time, the thermistor has important biological applications.

9.14 Heating Effects of Currents: Joule's Law

Consider a conductor which has a current I flowing through it (Fig. 9.10a). In the time interval dt, a quantity of charge dq, given by

$$dq = Idt$$

enters the portion of the circuit at terminal A and in the same time an equal quantity of charge leaves through B. As there is a transfer of charge dq from potential V_a to V_b, work dW must be done, given by

$$dW = dq\,(V_a - V_b) \qquad 9.10$$

$$= Idt\,V_{ab} \qquad 9.11$$

The power input, or rate of doing work, is given by

$$P = dW/dt$$

$$= IV_{ab} = I^2R \qquad 9.12$$

where R is the resistance between A and B.

Equation 9.12 is called *Joule's Law* and the power input is mainly dissipated in the form of heat.

What are the units of power in this case?

$$P = I \times V = \mathrm{C\,s^{-1}} \times \mathrm{J\,C^{-1}}$$

$$= \mathrm{J\,s^{-1}} = \text{Watts}$$

9.15 Circuit Equations

Joule's Law can be used to derive certain equations that are useful in solving problems on electrical circuits.

In Fig. 9.11 points a and b are at the same potential as they are connected together by a non-resistive wire. However, a potential difference exists across the resistor R. Suppose the current flowing round the circuit is I, then heat is developed in the external resistor at a rate I^2R. It is also found that every source of EMF has an internal resistance, usually low, called r. Hence the rate of heat development in the cell is I^2r.

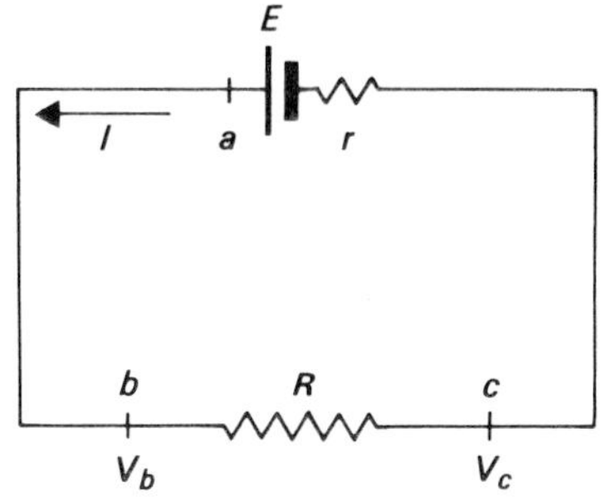

Fig. 9.11. An EMF E of internal resistance r is driving current through the external resistor R.

The total power supplied by the EMF is EI and this must be equal to the heat energy dissipated in the resistors

$$\therefore EI = RI^2 + rI^2 \qquad 9.13$$

or

$$I = \frac{E}{R + r}$$

Generally

$$I = \frac{\Sigma E}{\Sigma R} \qquad 9.14$$

Note that a convention must be adopted for the direction of current flow and the usual one is that current is the flow of the positive charge and so travels from the positive terminal of a source of EMF, round the external circuit, to the negative terminal.

9.16 Potential Difference between Points in a Circuit

The rate at which circulating charge gives up energy to the portion of the circuit between a and b (Fig. 9.12) is IV_{ab}. This is the power input to this portion of the circuit supplied by sources of EMF in the remainder of the circuit (not shown). Power is also supplied *by* the first source of EMF (E) and power is supplied *to* the second source (E'). Heat is developed in the various resistors.

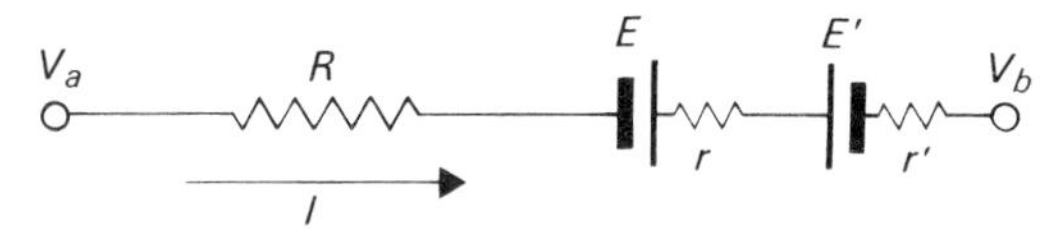

Fig. 9.12. (See text for explanation)

Equating power input and output

$$IV_{ab} + EI = E'I + (R + r + r')I^2 \quad 9.15$$

Equation 9.15 can be stated in a more general form

$$V_{ab} = \sum RI - \sum E \quad 9.16$$

Careful attention must be paid to algebraic signs. Direction from a to b is always considered positive. Currents and EMFs are positive if their direction is from a to b. Resistances are always positive.

If a and b coincide, $V_{ab} = 0$

$$\therefore I = \frac{\Sigma E}{\Sigma R} \quad 9.14$$

9.17 Series and Parallel Connection of Resistors

(i) Series (Fig. 9.13a)

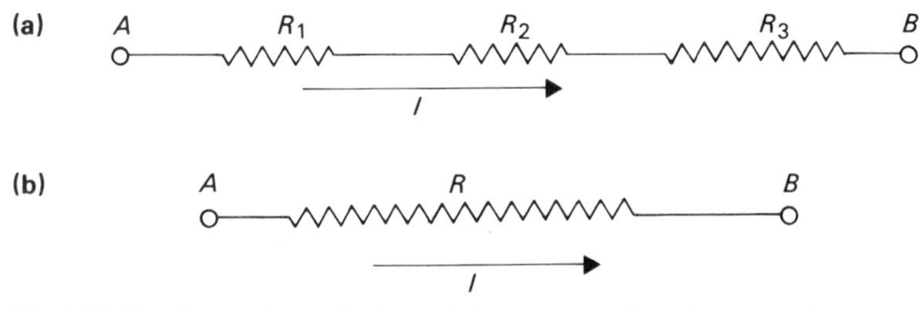

Fig. 9.13. The three resistors R_1, R_2 and R_3 connected in series (a) can be represented by the equivalent resistor R (b). Where $R = R_1 + R_2 + R_3$.

When the resistors are connected in series, the current through each is the same; hence

$$V_{ab} = IR_1 + IR_2 + IR_3$$

From Fig. 9.14b,

$$V_{ab} = IR$$

where R is the equivalent resistance. Hence

$$R = R_1 + R_2 + R_3 \qquad 9.17$$

(ii) Parallel

(a)

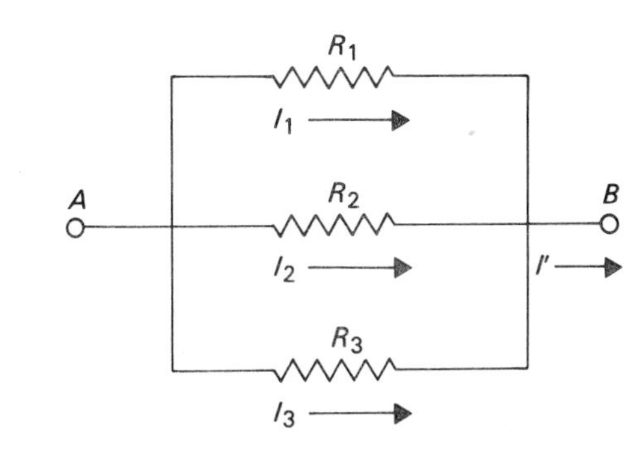

(b)

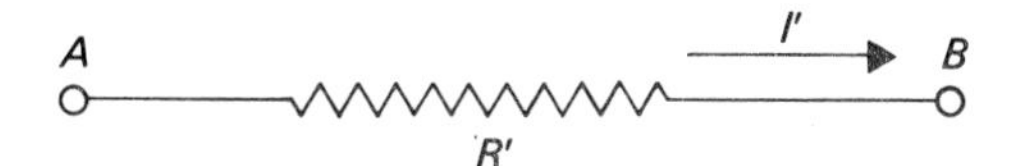

Fig. 9.14. The three resistors in parallel (a) can be represented by the equivalent resistor R' (b) where $1/R' = 1/R_1 + 1/R_2 + 1/R_3$.

When they are connected in parallel I' is split up into 3 paths, i.e.

$$I' = I_1 + I_2 + I_3$$

but the potential difference across the resistors is the same V_{ab} as they have common connections. Now

$$I_1 = \frac{V_{ab}}{R_1}, \qquad I_2 = \frac{V_{ab}}{R_2} \qquad \text{and} \qquad I_3 = \frac{V_{ab}}{R_3}$$

but from Fig. 9.14b

$$I' = \frac{V_{ab}}{R'}$$

$$\therefore \qquad \frac{V}{R'} = \frac{V}{R_1} + \frac{V}{R_2} + \frac{V}{R_3}$$

so

$$\frac{1}{R'} = \frac{1}{R_1} + \frac{1}{R_2} + \frac{1}{R_3} \qquad 9.18$$

9.18 Kirchoff's Rules

Not all networks can be reduced to simple series or parallel arrangements (Fig. 9.15) and in order to analyse such networks another approach, devised by Gustav Kirchoff, has to be used. He introduced the terms *branch point* and *loop* and they are defined as follows:

(i) A *branch point* occurs when three or more conducting paths meet, e.g. points *a* and *b* in Fig. 9.15.

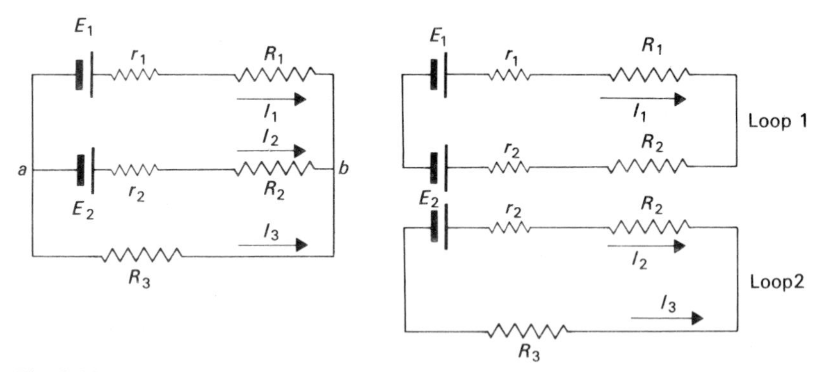

Fig. 9.15. Example of a network that cannot easily be analysed in terms of series and parallel arrangements of resistors. *a* and *b* are the two branch points and loop 1 and loop 2 are the two loops.

(ii) A *loop* is any closed conducting path.

The basic rules for the solution of network problems are:

(i) *Point rule.* The algebraic sum of the currents *towards* any branch point is zero, i.e.

$$\sum I = 0 \qquad 9.19$$

Note that if there are n branch points in a network, there are only $n - 1$ independent point equations.

(ii) *Loop rule.* The algebraic sum of the EMFs in any loop equals the algebraic sum of the *IR* products in the same loop, i.e.

$$\sum E = \sum IR \qquad 9.20$$

Some sign convention must also be adopted in conjunction with the rules and a useful one is:

(i) *Point rule convention.* Current is considered positive if its direction is towards a branch point.

(ii) *Loop rule convention.* The clockwise direction is considered positive and all currents and EMFs in this direction are positive.

Problem 9.1

Write down the equations to solve the network in Fig. 9.15.

Answer

There are two branch points, *a* and *b* and so there is only one independent point equation.

$$I_1 + I_2 + I_3 = 0$$

considering loop 1

$$E_1 - E_2 = I_1 r_1 + I_1 R_1 - I_2 r_2 - I_2 R_2$$

and loop 2

$$E_2 = I_2 r_2 + I_2 R_2 - I_3 R_3$$

There are now three independent equations to solve for the three unknown currents.

Problem 9.2

A Wheatstone Bridge (Fig. 9.16) is an instrument for determining the value of an unknown resistor. It consists of 4 resistors, a source of EMF, and a current detector (*ammeter*) arranged as shown. The known variable resistor R_2 is adjusted until there

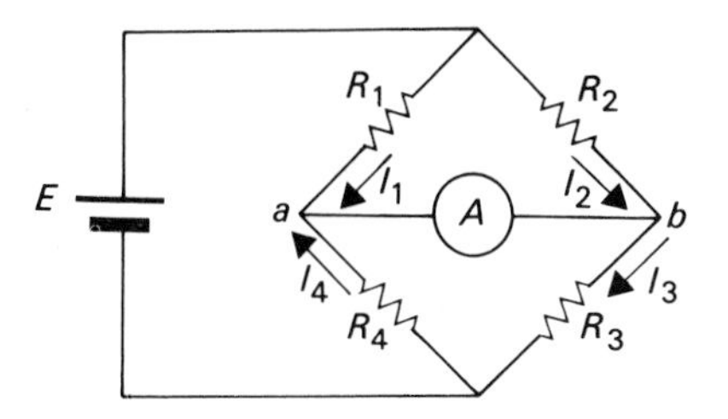

Fig. 9.16. The Wheatstone Bridge. R_3 and R_4 are fixed, known resistors; R_2 is known and variable; R_1 is the unknown resistor. When the bridge is in balance there is no current flowing through the ammeter *A*.

is a zero reading on the ammeter and the bridge is then said to be in balance. The reader should easily be able to show that the magnitude of the unknown resistor, R_1 is given by

$$R_1 = R_2(R_4/R_3)$$

where R_3 and R_4 are known resistors of fixed value.

9.19 Membranes – Potentials and Conductances

Membrane potentials are set up by the diffusion of ions down their *electrochemical gradients.* Although the actual mechanism of the passage of ions through membranes is not understood, we can obtain some understanding of the problem by the application of elementary thermodynamics.

Suppose we have a membrane separating two KCl solutions of different concentrations ($K_i > K_0$) and suppose the membrane is permeable only to potassium ions (Fig. 9.17). The potassium ions will diffuse down their concentration gradient, taking positive charge with them so that the outside phase will acquire a positive potential with respect to the inside. This potential will tend to impede the subsequent movement of cations and a point will be reached when there will no longer be a net diffusion of potassium ions from inside to out. At this point the driving force on the ions due to the concentration gradient will be exactly balanced by the driving force due to the potential difference.

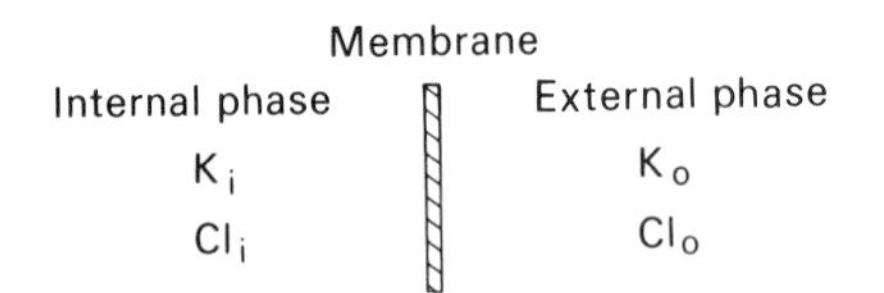

Fig. 9.17. A membrane separating two phases. It is assumed that only potassium ions can pass through.

We can obtain the magnitude of this potential, called the *equilibrium* or *Nernst potential,* by considering the work that has to be done to move a small quantity of K^+ ions from the outside phase to the inside. From elementary thermodynamics (Spanner, 1964) the work required to move δn moles through the membrane against the concentration gradient is given by

$$\delta W = \delta n R T \log_e \frac{[K_i]}{[K_0]} \qquad 9.21$$

where R is the gas constant; T the absolute temperature; and K_i and K_0 the concentrations of potassium in the internal and external phases respectively.

Now the work done to move δn moles against a potential difference E is

$$\delta W = \delta n Z F E \qquad 9.22$$

where Z is the valency of the ion (equiv. mol^{-1}); F is the Faraday (96,500 C per equiv.); and E is the potential difference across the membrane. The reference for potential is taken as the external solution.

At equilibrium, there will be no net work done, and so

$$\delta n Z F E = -\delta n\, R T \log_e \frac{[K_i]}{[K_0]}$$

and $$E = -\frac{RT}{ZF} \log_e \frac{[K_i]}{[K_o]} \qquad 2.23$$

$$= -25 \log_e \frac{[K_i]}{[K_o]} \times 10^{-3}\,\text{V at } 20^\circ\text{C}$$

$$= -58 \log_{10} \frac{[K_i]}{[K_o]} \times 10^{-3}\ V$$

The actual membrane potential difference in a biological system will contain contributions from all of the ionic species present, and the contribution from each will depend on the *permeability* of the species. The more permeable the ion, the more it will contribute to the potential. In most animal cells, for example, chloride and potassium are much more permeable than sodium, so they contribute most to the potential.

In analysing complex electrical currents in the membrane it is often helpful to have a model to work from and membrane biophysicists make great use of the equivalent circuit (Fig. 9.18). The diffusion potentials are represented by batteries and the

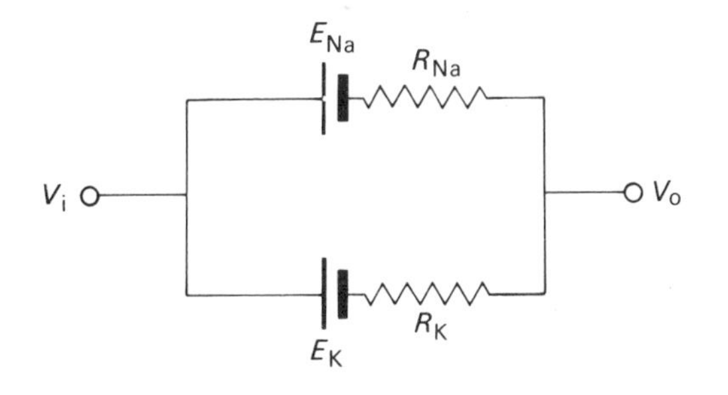

Fig. 9.18. A simple equivalent circuit of a membrane. R_{Na} and R_K are the resistances of the sodium and potassium channels and E_{Na} and E_K are the equilibrium (Nernst) potentials for sodium and potassium.

actual contribution which these make to the overall membrane potential will be determined by the ease with which ions can pass through their appropriate channel in the membrane. The lower the permeability of an ion, then the greater will be the resistance of that channel. It is left as an exercise to show that when there is no net current across the membrane, the potential across it is given by:

$$V_i - V_o = \frac{E_{Na} R_K}{R_{Na} + R_K} + \frac{E_K R_{Na}}{R_{Na} + R_K} \qquad 9.24$$

The resistance of the cell membrane is normally measured by passing a current pulse ΔI across the membrane via two electrodes, one internal and one external, and measuring the resultant change in potential ΔV across another pair. If the area of cell membrane is A, then the current through a unit area is $\Delta I/A$ and the resistance is the change in potential divided by the current, i.e.

$$R_m = A \frac{\Delta V}{\Delta I} \qquad 9.25$$

Resistance in this case therefore has the units of Ωm^2 and the resistance of most animal cell membranes is in the region of $10^{-1}\,\Omega m^2$. The conductance is therefore $10\Omega^{-1}\,m^{-2}$.

Problem 9.3

The resistance of most cell membranes is of the order $10^3\ \Omega\ cm^2$ ($10^{-1}\ m^2$). We wish to measure the transmembrane potential in an amphibian oocyte of diameter 10^{-3} m and in order to do this we have to insert a microelectrode with a tip resistance of $10^6\ \Omega$. The tip resistance of the reference microelectrode is also $10^6\ \Omega$. Show that a *voltmeter* with a high internal resistance, R_v, has to be used for an accurate measurement of the potential.

The circuit for the system is given in Fig. 9.19.

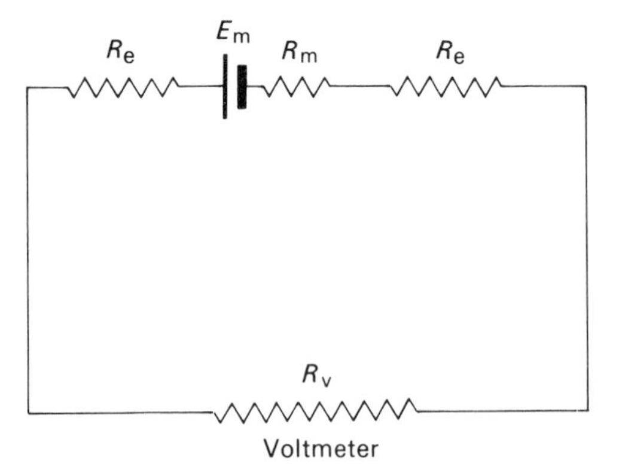

Fig. 9.19. The membrane of the oocyte is represented by a battery in series with a resistance, E_m and R_m. The reading on the voltmeter is V and you have to show that $V \approx -E_m$ only when $R_v \gg R_m + 2R_e$, where R_e is the tip resistance of a microelectrode.

9.20 Capacitance

The discussion so far has dealt with steady electrical currents. When the voltage is not constant, as in the case when it is first applied, or if it is an alternating voltage, e.g. in power lines, then some additional phenomena are seen.

For any current to flow in a circuit there must be a complete path of finite resistance through which the electrons can flow and return to their starting place. However, when a voltage is first applied to a conductor there will be a movement of electrons into the conductor even in the case of an open circuit. What happens then is that the excess of electrons in the conductor produces an electric charge which builds up with still more electrons until it balances the applied voltage. Then with no voltage gradient, there will be no further movement of electrons, and hence no current. The situation described could be called a one-plate capacitor (Fig. 9.20).

If two sheets of metal are placed close together (Fig. 9.21), but separated from contact by a good insulator, the same movement of electrons will occur when a voltage is applied across the two plates. However, on one side of the circuit electrons will flow out of the plate leaving a net positive charge, while on the other, electrons will flow in and the whole applied potential will be concentrated in a rather steep gradient across the insulator between the plates. This insulator, called the *dielectric,* can by its dimensions and dielectric properties, affect the ultimate density of electrons on the plates required to equalise the applied voltage. Thus the charge on the capacitor,

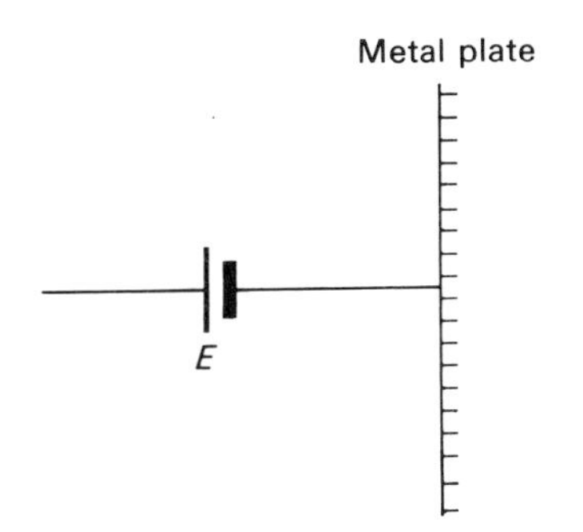

Fig. 9.20. Current only flows for a short time when the battery is connected to the metal plate, because an opposing voltage is set up equal in magnitude but opposite in sign to the EMF.

which means the actual number of electrons in excess or deficit on the plates, varies not only as the applied voltage, but also with the area of the plates and the type of dielectric between the plates. It can be shown that the capacitance of a parallel plate condenser is given by

$$C = \frac{K \epsilon_0 A}{d} \qquad 9.26$$

where K is the relative dielectric constant; ϵ_0 is the absolute permittivity; A is the area of the plates; and d the distance between them.

The term *capacitance* denotes a capacity to store electrons and the relationship between stored charge and applied potential is given by

$$Q = CV \qquad 9.27$$

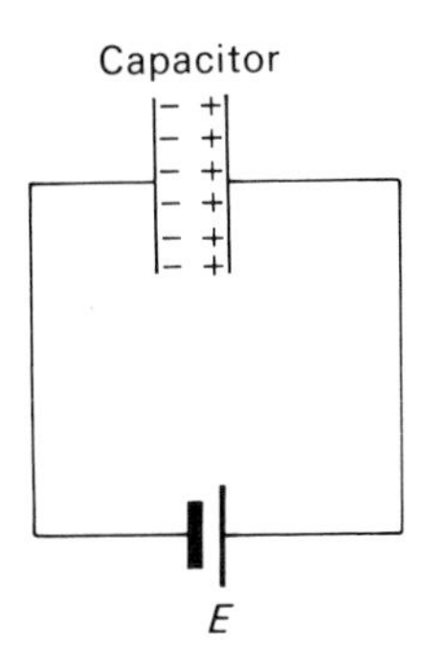

Fig. 9.21. (See text for explanation)

When Q is in coulombs and V is in volts, the units of C are *farads.* A capacitor of one farad will store enough electrons for a one-volt applied potential to require a net charge of one coulomb to counteract the applied potential. This is a very large number of electrons and capacitors in the microfarad and picofarad range are more commonly found in biology. The capacitance of most cell membranes is of the order of 10 mF m^{-2}.

In a steady direct current situation, once the capacitor is charged to the maximum for the applied potential, no more current will flow and the circuit resembles an open circuit. When the current is reversing its direction rapidly, as in alternating current circuits, it is obvious that the electrons will be able to move back and forth from one plate of the condenser to the other, with each change in polarity of the applied voltage, so there will be a current flow in the circuit, even though there is an insulating dielectric between the plates. If the alternations are slow, then there will be time for an equilibration to be reached at each point in the cycle and the number of electrons on the plates will create a counter potential which will stop the flow of current. However, if the voltage reverses very rapidly and, if the electrical resistance is high enough to limit the current flow, there will never be enough electrons on the plate of the condenser to effectively counteract the applied potential. In such a case the current is limited by the resistence in the circuit alone. Thus the limiting effect of the condenser on the current depends both on its capacitance C and also on the frequency f of alternations of the voltage. The limiting effect is called the *capacitive reactance* X_c, and is measured in ohms, just as resistance is

$$X_c = \frac{1}{2\pi fC} \qquad 9.28$$

A capacitor is usually thought of in terms of metal plates separated by an insulating medium, dielectric. The concept is however equally applicable to the cell membranes where $C \approx 10\ mFm^{-2}$, or $1\mu F\ cm^{-2}$ using the units most likely to be met.

9.21 Energy Stored in a Capacitor

As charging a capacitor involves the movement of charge from the plate at a lower potential to the plate at a higher potential, work must be done.

Suppose that a total of q units of charge have been added to one plate and that the potential difference between the plates is V, then the work done dW in transferring further charge dq will be given by

$$dW = V\,dq \qquad 9.29$$

and as

$$q = C\,V \qquad 9.27$$

$$dW = \frac{q}{C}\,dq$$

The total work W in increasing the charge from 0 to Q is

$$W = \int_o^Q \frac{1}{C}\,q\,dq \qquad 9.30$$

$$= \frac{1}{2}\frac{Q^2}{C} = \frac{1}{2}CV^2 \qquad 9.31$$

This work done is stored as energy in the capacitor and is released when it discharges.

9.22 Dielectric Constant

The dielectric constant is an important property of matter and the value for many materials can readily be determined by sandwiching them between the plates of a capacitor. The capacitance of this system is then measured by a current pulse technique (section 9.23) and the dielectric constant can be obtained from equation 9.26 if the geometry of the capacitor (A/d) is known.

K, known as the relative dielectric constant, is nearly 1 for most gases, but for some substances which are strong dipoles the value is much greater, e.g. for water, $K = 80.4$ (20°C). The physical basis of the high dielectric constant of dipolar media is that when the potential difference is applied, the dipoles rotate, aligning themselves with the electrical field (Fig. 9.22). The orientated dipoles act to neutralize the charge on the plates thus increasing the capacity of the system. The bulk of the energy stored in the capacitor is the energy required to orientate the dipoles from the random state. If the field is allowed to decay, the dipoles randomize, and providing there is no hysteresis, the energy is recovered. This randomization or relaxation of the dipoles is effectively a flow of current across the system.

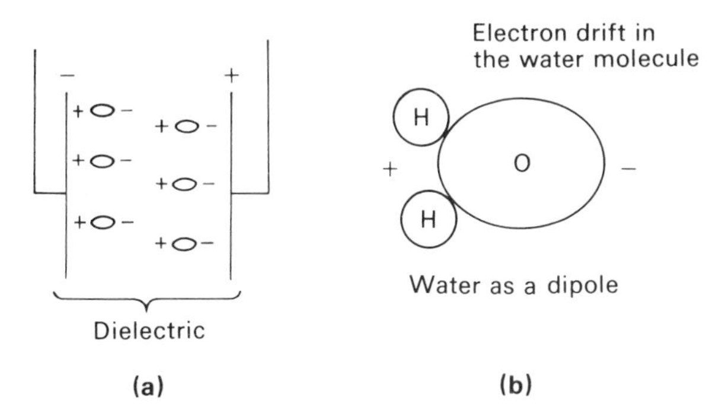

Fig. 9.22. (See text for explanation)

In the case of the cell membrane, the capacitance per unit area ($10 \mathrm{mFm^{-2}}$) defines the ratio $K\epsilon_0/d$. If $d \approx 5$ nm, then $K \approx 5$. A value as low as this might be regarded as surprising for a system mainly composed of lecithin, which is a strong dipole. However, there is evidence that the lecithin bilayer has a considerable degree of order, although it would be incorrect to regard it as in a frozen, orientated state. The term *liquid crystal* is used for such a state as the cell membrane. The degree of order restricting the free rotation of the dipoles would be responsible for a low dielectric constant in such a bilayer.

9.23 Circuits containing Capacitance and Resistance (Fig. 9.23)

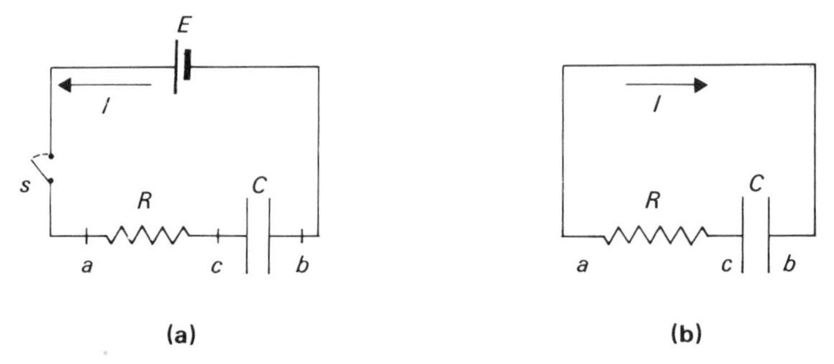

Fig. 9.23. (a) When the switch S is closed, a current I will flow round the circuit to charge up the capacitor. (b) After charging up, if the ends ab are short-circuited, current will flow and the capacitor will be discharged.

First let us derive a very useful relationship between current, capacitance and potential difference, starting with equation 9.27:

$$Q = CV$$

and on differentiating,

$$dQ/dt = CdV/dt$$

as capacitance is independent of time. Hence

$$I = CdV/dt \qquad 9.32$$

When the switch S is closed a current I will flow to charge up the condenser

$$E = V_{ac} + V_{cb}$$

$\therefore$ $$V_{cb} = E - IR$$

but $$I = C\,dV_{cb}/dt$$

$\therefore$ $$V = E - RCdV/dt$$

V_{cb} will simply be referred to as V

$$R\,C\,dV/dt = E - V$$

$$\frac{dV}{E-V} = \frac{1}{RC}dt$$

and on integration

$$-\ln(E-V) = \frac{1}{RC}t + K$$

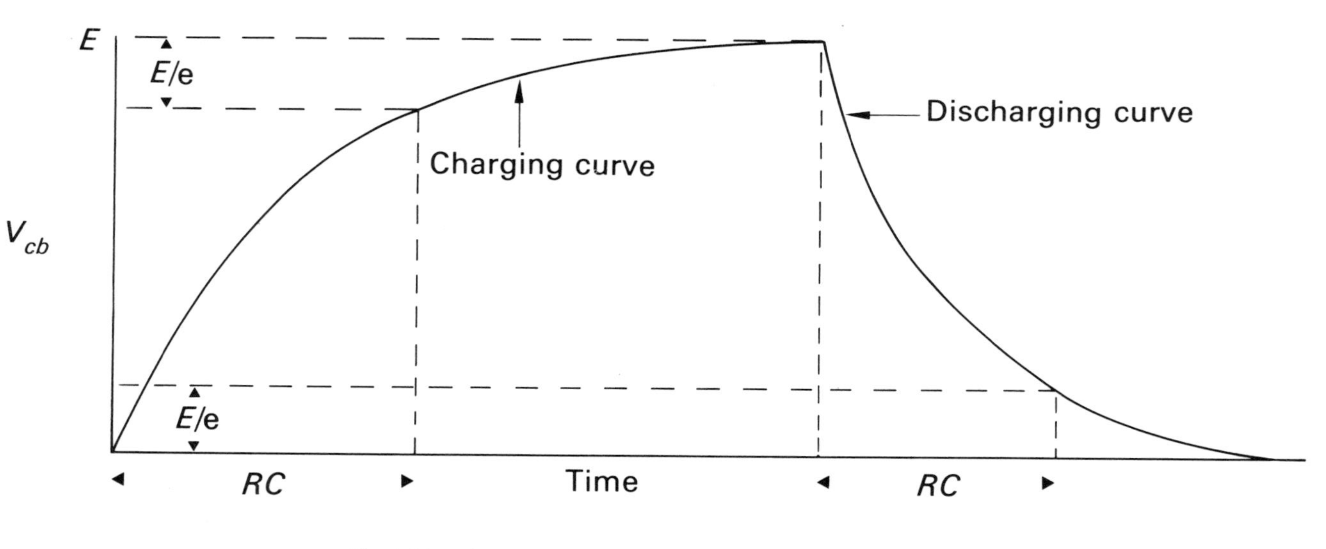

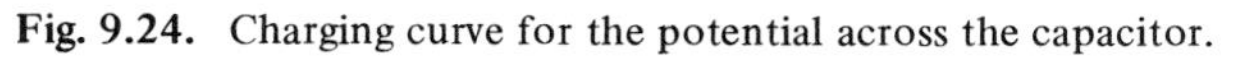

Fig. 9.24. Charging curve for the potential across the capacitor.

where K is the integration constant. Or

$$(E - V) = \exp\left(-\frac{t}{RC}\right)\exp(-K)$$

now when $t = 0$, $V = 0$

$\therefore \quad \exp(-K) = E$

i.e. $\quad E - V = E\exp(-t/RC) \qquad 9.33$

$\therefore \quad V = E(1 - \exp t/RC)$

RC (Fig. 9.24) is referred to as the *time constant* of the circuit and has units of seconds.

Now suppose that after charging up the condenser to the potential E, we let it discharge through the resistance R, i.e. disconnect the source of EMF and connect the points a and b together (Fig. 9.23b)

$$V_{ac} + V_{cb} = 0$$

i.e. $\quad -IR + V_{cb} = 0$

where current is flowing from c to a across the resistor, and

$$I = -C\frac{dV_{cb}}{dt}$$

as the capacitor is discharging, therefore

$$V = -RC\frac{dV}{dt}$$

writing V for V_{cb} again.

If we integrate once more and take the boundary condition $V = E$ when $t = 0$, the solution to this equation is

$$V = E\exp(-t/RC) \qquad 9.34$$

In this case the time constant RC is the time for the potential to decrease to $1/e$ of its original value (Fig. 9.24). The capacitance of a cell membrane, for example, can be determined by passing a square pulse of current (Fig. 9.25a) through the membrane by means of a pair of glass microelectrodes. The potential difference across the membrane is measured simultaneously by means of a further electrode pair and the voltage response shows typical charging and discharging characteristics. As the membrane resistance is given by the ratio V_{max}/I, the membrane capacitance (mFm^{-2}) can be calculated from the RC characteristic of the membrane.

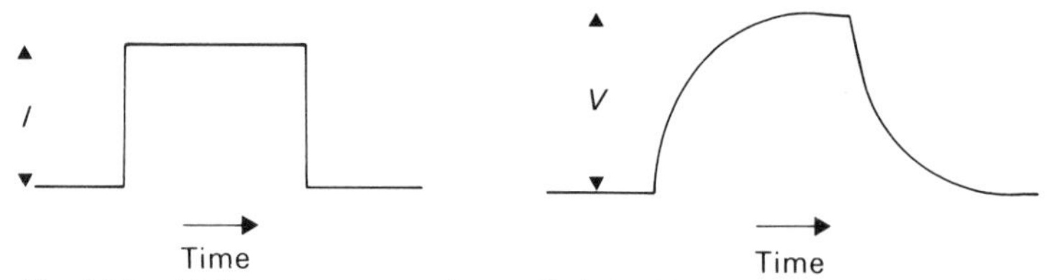

Fig. 9.25. (a) Square current pulse applied. (b) Voltage response observed.

9.24 Capacitors in Series and Parallel

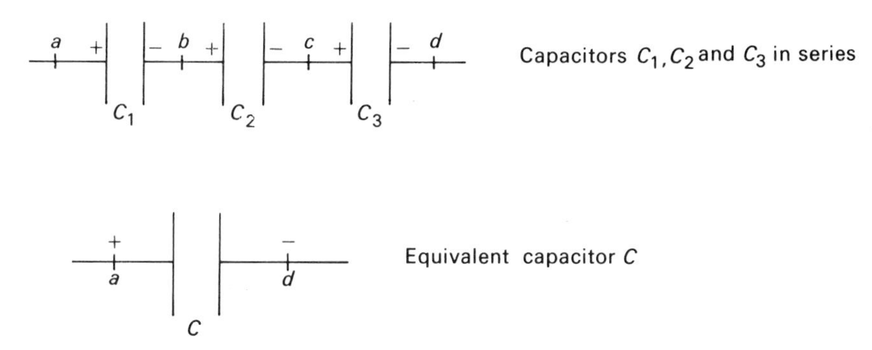

Fig. 9.26. (a) Capacitors C_1, C_2 and C_3 in series. (b) Equivalent capacitor C.

(i) Capacitors in series (Fig. 9.26)
If the left hand plate of C_1 receives a charge $+Q$, then a charge $-Q$ is induced on the right hand plate and $+Q$ appears on the left hand plate of C_2 etc.

$$Q = C_1 V_{ab} = C_2 V_{bc} = C_3 V_{cd}$$

and $$V_{ad} = V_{ab} + V_{bc} + V_{cd}$$

and if C represents the equivalent capacitance, i.e. the capacitance of the single capacitor that would become charged with the same charge Q when the p.d. across its terminals is V_{ad}.

Then

$$V_{ad} = \frac{Q}{C}$$

$$\therefore \quad \frac{Q}{C} = \frac{Q}{C_1} + \frac{Q}{C_2} + \frac{Q}{C_3}$$

$$\therefore \quad \frac{1}{C} = \frac{1}{C_1} + \frac{1}{C_2} + \frac{1}{C_3} \qquad 9.35$$

(ii) Capacitors in parallel (Fig. 9.27)
The potential across each capacitance is V_{ab}, but the charge induced on each is different.

$$Q_1 = C_1 V, \qquad Q_2 = C_2 V, \qquad Q_3 = C_3 V$$

The total charge on the parallel network is

$$Q = Q_1 + Q_2 + Q_3$$

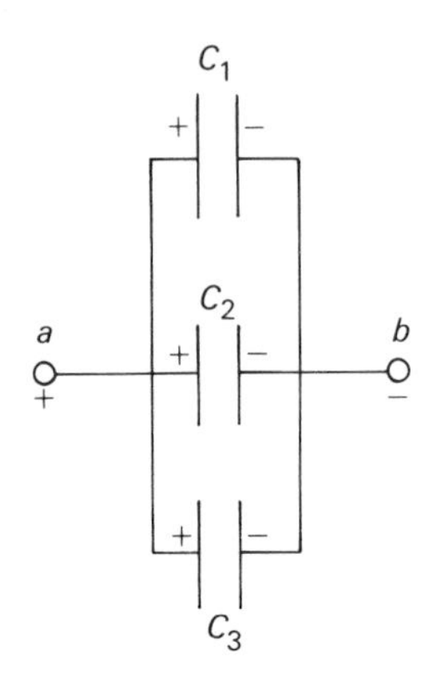

Capacitors C_1, C_2 and C_3 in parallel

Fig. 9.27. (See text for explanation)

Defining the equivalent capacitance as the one for which

$$Q = C\,V_{ab}$$

$$C\,V_{ab} = C_1\,V_{ab} + C_2\,V_{ab} + C_3\,V_{ab}$$

$$\therefore \quad C = C_1 + C_2 + C_3 \qquad 9.36$$

Compare the equivalent capacitors in (a) and (b) with the equivalent resistors for the series and parallel arrangements (section 9.17).

Problem 9.4

A membrane can be represented by the equivalent circuit shown in Fig. 9.28. E_{Na} and E_K are the sodium and potassium Nernst potentials and R_{Na} and R_K represent the resistances of the sodium and potassium channels in the membrane. C is the membrane capacitance and E_m, the transmembrane potential is given by $V_i - V_o$.

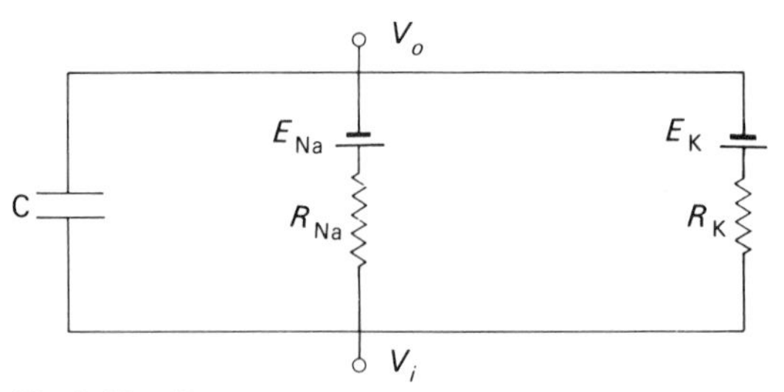

Fig. 9.28. (See text for explanation)

(*a*) The followimg data has been obtained for squid nerve in the resting state. Na_o = 440 mM, K_o = 20 mM, Na_i = 50 mM, K_i = 400 mM; E_m = −60 mV (inside negative)

and total membrane resistance equals 1 kΩ cm². Show that E_{Na} = +55 mV, E_K = −75.5 mV and calculate the ratio of the resistances of the sodium and potassium channels.
(*b*) At the height of the action potential, E_m = +40 mV. Calculate the total resistance at the height of the action potential, assuming that only R_{Na} has had time to change and compare your computation with the experimental value of 40 Ω cm². Why is the computed resistance the larger?
(*c*) The membrane capacitance is 1 μF cm^{-2}. How many moles of ions move through the membrane in order to discharge it during an action potential? Take ZF = 96,500 coulomb/mole. See also Duncan and Croghan (1973) for an equivalent circuit analysis of photoreceptor potentials (Fig. 9.6).

9.25 Magnetism and Electromagnetism

These complex phenomena have, at least at present, little direct relevance to us as, apart from the possible role played by magnetic field detectors in bird navigation, few biological processes are directly influenced by magnetic fields. Indirectly, however, there are occasions when the biologist has to concern himself with magnetism. For example in the study of nuclear magnetic resonance spectra **(NMR)**, a strong magnet can be used to line up the randomly oriented nuclear magnetic moments in a population of say ^{23}Na ions. By studying the time course of the return of the nuclear magnets to the random state after the removal of the magnetic field information can be obtained about the interaction of sodium ions both with one another and with molecules in the surrounding medium. Studies such as these are not limited to sodium ions, but can be prosecuted with any atom that has an unpaired nuclear magnetic moment.

The essential difference between an unmagnetized and magnetized metal bar is that the former has a random orientation of the *magnetic domains* making up the bar, while in the latter they are lined up (Fig. 9.29). The magnetized bar behaves as if it

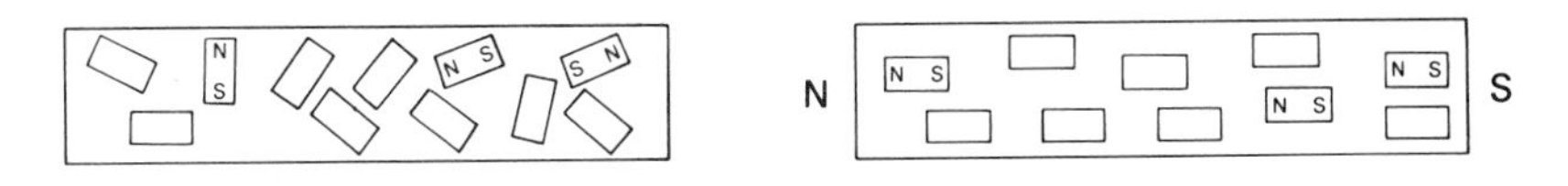

Fig. 9.29. The positions of the north and south poles in the hypothetical magnetic domains are denoted by N and S respectively.

were asymmetrical in some way – it is said to have two poles, one north and the other south. When the two north poles from two magnets are brought together then the magnets repel one another. A magnet is said to have a magnetic field associated with it which modifies in some way the space surrounding the magnet. This field will interact with the field from another magnet if the two are brought together and a repulsive or attractive force will result.

The magnetic field of a bar magnet is shown in Fig. 9.30a. The density of the field lines, magnetic flux density, is a measure of the intensity of the magnetic field. Soft iron bars tend to concentrate field lines (Fig. 9.30b) so the flux density inside the iron bar is greater than it would be if the bar was replaced by air or glass.

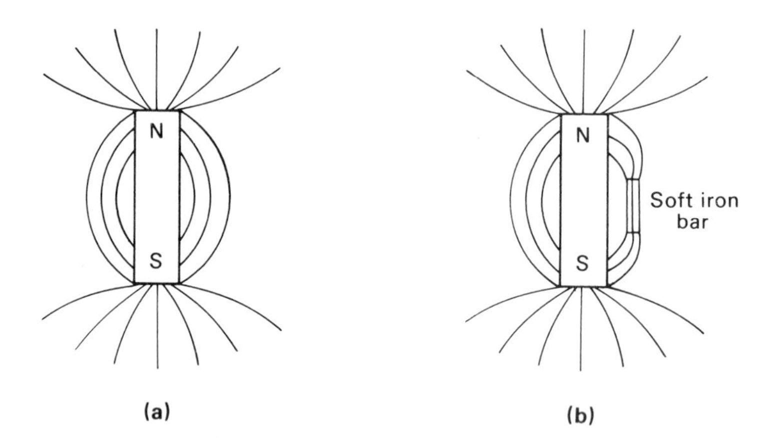

Fig. 9.30. (See text for explanation)

Magnetic fields are also set up by currents flowing along conductors and this can by clearly seen if a compass needle is placed near a wire carrying a direct current (Fig. 9.31).

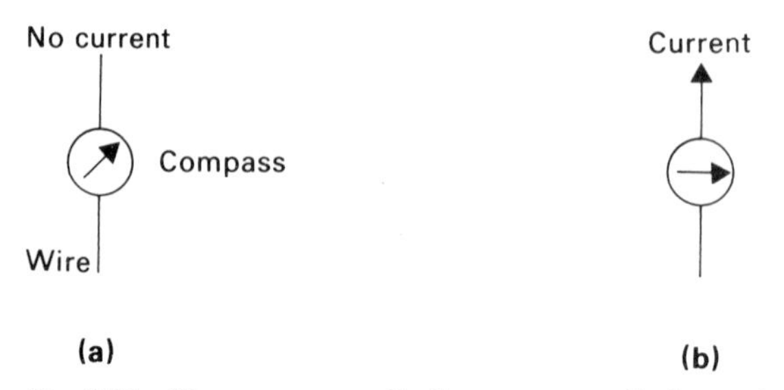

Fig. 9.31. The compass needle lines up perpendicular to the wire when current is passing (b).

Two wires a short distance apart exert forces on one another as a result of the interaction of the fields (Fig. 9.32). If the currents are flowing in the same direction, the force is an attractive one and it is repulsive if the currents are moving in opposite directions. The forces depend on the currents flowing in the wires and the distance apart of the wires and as the force can be very easily measured, the arrangement is used to define the *ampere.*

The ampere (amp) is that intensity of current which flowing in each of two long parallel conductors one metre apart, in vacuo, results in a force of 2×10^7 Nm^{-1} between the conductors. This also serves to define the coulomb as the charge transferred by a current of one ampere flowing for one second.

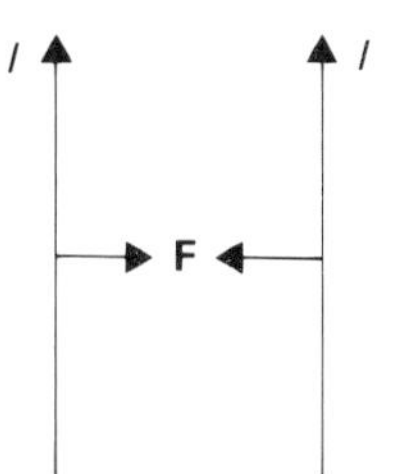

Fig. 9.32. Two wires with current flowing in the same direction experience an attractive force F.

9.26 Magnetic Fields

Magnetic fields exert forces on current carrying conductors (Fig. 9.33) and the vector relationship between the two is given by

$$\mathbf{F} = I\,\mathbf{L} \times \mathbf{B} \qquad 9.37$$

where **B** is the magnetic field; I is the current in the conductor, and **L** is the length of the conductor. This equation serves as a definition of the units of magnetic field. One *Weber* m^{-2} is defined as that magnetic field which will exert a force of 1 N on a 1 m

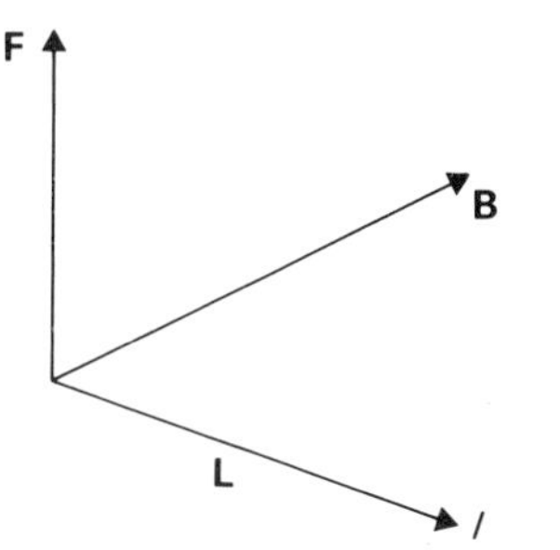

Fig. 9.33. A conductor carrying a current I in a magnetic field B experiences a force F which acts at right angles to both L and B.

length of wire carrying a current of 1 A and lying in a direction at right angles to the magnetic field. The converse phenomenon is also true. If a conductor is moved in a magnetic field, the charges within it experience a force. The movement of these charges constitutes a flow of current, which in turn produces a force tending to oppose the movement of the conductor. Work will then have to be done to move the conductor. Suppose the conductor moves a small distance $d\mathbf{x}$, then the work done, dW is given by

$$dW = \mathbf{F} \cdot d\mathbf{x} = I\,\mathbf{L} \times \mathbf{B} \,.\, d\mathbf{x} \qquad 9.38$$

The power input is dW/dt, and therefore

$$\text{power} = I\,\mathbf{L} \times \mathbf{B} \cdot \mathbf{v} \qquad 9.39$$

where **v** is the velocity of the conductor.

But the power input is also given by IV (equation 9.12)

Therefore

$$V = \mathbf{L} \times \mathbf{B} \cdot \mathbf{v} \qquad 9.40$$

V is the p.d. across the ends of the conductor and **v** is the velocity of the conductor. B is the flux density of magnetic field in Wb m^{-2}.

When **L**, **B** and **v** are mutually perpendicular, as in the case of the electromagnetic flow meter (section 9.27), equation 9.40 simplifies to

$$V = \mathrm{L\,B\,v} \qquad 9.41$$

9.27 Electromagnetic Flow Meter

The electromagnetic blood flow transducer (Fig. 9.34) consists of an electromagnet to generate a magnetic field and two electrodes to sense the flow signal (V_f). They are

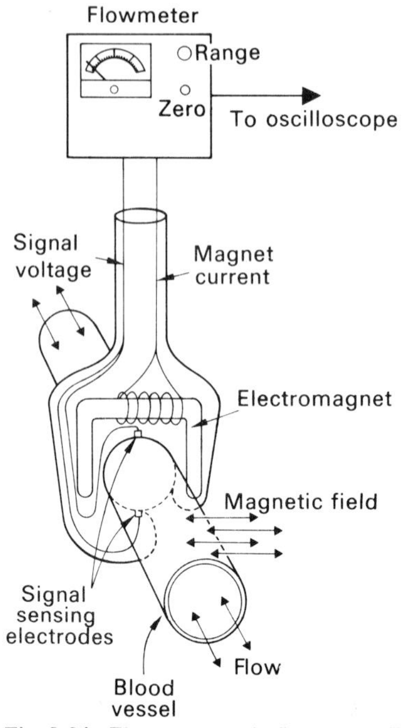

Fig. 9.34. Electromagnetic flowmeter. Equation 9.41 can be applied directly to this system as the blood flow, i.e. velocity of the conductor, the conductor length (distance between sensing electrodes), and the magnetic field are all at right angles to one another. (From Strong, 1970). (Reproduced by permission of Tektronix Inc.)

encapsulated in an inert hard plastic in a form which permits them to fit around the blood vessel of interest. The lumen or inside diameter of the holder slightly deforms the vessel so that its cross-sectional area is now fixed and indeed known. In this way the transducer can be used to measure the flow J, i.e. $V_f \alpha J$, although basically it is a mean velocity transducer (i.e. $V_f \alpha$ v).

Problem 9.5

An electromagnetic flowmeter with a sensing head of cross sectional area 0.07×10^{-4} m^2 and a fixed distance between sensing electrodes of 1×10^{-3} m is used to measure blood flow in an artery. When a magnetic field of 2Wb m^{-2} was applied, a peak signal of 600 μV was observed on the oscilloscope screen. Calculate the blood flow (m^3 s^{-1}) through the artery.

References

Aidley D.J. (1971) *The Physiology of Excitable Cells.* Cambridge University Press.

Donaldson P.E.K. (1958) *Electronic Apparatus for Biological Research.* Butterworth, London.

Duncan G. & Croghan P.C. (1973) Excitation and Adaptation in the Cephalopod Retina: an equivalent circuit model. In *Biochemistry and Physiology of Visual Pigments.* ed. H. Langer, Springer-Verlag, Berlin.

Hodgkin A.L. (1964) *The Conduction of the Nervous Impulse.* Liverpool University Press.

Jarman M. (1970) *Examples in Quantitative Zoology.* Arnold, London.

Katz B. (1966) *Nerve, Muscle and Synapse.* McGraw-Hill, New York.

Lissman H.W. (1963) Electric Location by Fishes. In *From Cell to Organism,* Freeman, San Francisco.

Sears F.W. & Zemansky M.W. (1964) *University Physics.* Addison-Wesley, Reading, Mass.

Strong P. (1970) *Biophysical Measurements.* Tektronix, Beaverton.

Worcester R. (1969) *Electronics.* Hamlyn, London.

Young S. (1973) *Electronics in the Life Sciences.* Macmillan, London.

Chapter 10
Radiations and Radiobiology

10.1 Introduction

Two discoveries made at the end of the 19th century in the field of atomic physics have profoundly influenced the progress of modern biology. In 1895, Röntgen, while working with an early form of the cathode ray tube, found that rays from it caused fogging of photographic plates and he suggested that unknown, invisible radiations or X-rays were being produced by the tube. He also took the first X-ray photograph by placing his hand between the tube and a photographic plate.

In the following year Becquerel discovered *natural radioactivity* and showed that uranium ores placed near photographic plates caused fogging in the dark. He was unfortunate to be one of the first scientists to suffer from radiation as he used to carry his ores around in his pocket.

We have previously seen that X-rays are highly energetic photons released from excited atoms during electronic transitions between the innermost orbitals (Chapter 8, p. 119); natural radioactivity on the other hand is a property of the atomic nucleus.

10.2 The Atomic Nucleus and the Radiation Spectrum

The nucleus consists of positively charged *protons* and uncharged *neutrons* held together by very strong nuclear forces set up by the exchange of elementary particles called *mesons* among the neutrons and protons. In so-called *radioactive* nuclei the forces are unable to overcome the mutual repulsion of the constituents, the nucleons, and the stable state is reached by a shedding of excess energy in the form of ionizing radiations. Radioactive atoms occur naturally, but they can also be produced artificially.

The first radiations to be characterized were the α, β, and γ-rays from uranium ores. *γ-rays* are highly energetic photons arising from a rearrangement of energy levels within the nucleus; *β-rays* are highly energetic electrons arising from a neutron to proton transition within the nucleus

$$n^0 \rightarrow p^+ + \beta^- \qquad 10.1$$

An additional elementary particle called the antineutrino should also appear on the right hand side of equation 10.1, but it has been omitted as it has no known biological effect.

Some radioactive atoms, e.g. uranium, emit the very large *α-particle* to reach a more stable state and in this case two protons and two neutrons are emitted as one entity. *Positrons* (positive electrons), *neutrons* and *protons* can also be produced from radioactive atoms. This chapter will deal with the effects of the radiations most likely to be encountered by the biologist. The biological effects of the multitude of elementary particles found only in the depth of space or at the centre of a nuclear reactor or particle accelerator will be left to the enthusiast to research and for those the book by Thornburn (1972) is a good starting point.

Radiations are partly characterized by the energy of the particle or photon emitted and this information enables anyone handling a radioactive source to take adequate precautions to shield it, because the range of the radiation depends on the energy. The data for energy and range in tissue for radiations that a biologist might encounter are given in Table 10.1.

Table 10.1. The range for the photons is the so-called half-thickness of tissue, i.e. the thickness that will reduce the initial intensity by one half. The range of the particles on the other hand is the thickness that will stop the particle. 1 keV = 10^3 eV and 1 MeV = 10^6 eV.

Radiation	Energy range	Range in tissue
α particle	1–20 MeV	1–10 μm
β particle	10 keV–15 MeV	10 μm–5 x 10^{-2} m
X and γ photons	10 keV–2 MeV	10^{-3} m–10^{-1} m

10.3 Sources of Radiation

Anyone who has had an X-ray taken in hospital or dentist's surgery will have been knowingly exposed to a certain amount of radiation. However, there are more and less obvious sources of radiations that are to be found all around us, and indeed in the very matter we are made of. These ubiquitous sources are the so-called *radioisotopes.*

Isotopes are atomic species that have the same number of protons in their nuclei, but different numbers of neutrons. They have the same number of electrons, so their chemical properties are identical, but because of the differing proportions of neutrons and protons, some of the nuclei will be more stable than others. The unstable isotopes are called *radioisotopes*, as in order to reach a more stable state they emit energy in the form of radiations.

An isotope of sodium, for example, that occurs naturally has eleven protons and twelve neutrons in its nucleus. The shorthand description of this isotope is $^{23}_{11}Na$, and sometimes only ^{23}Na. This isotope is stable. ^{24}Na on the other hand is a radio-isotope with eleven protons and thirteen neutrons and it is widely used in both biology and medicine. It is produced in nuclear reactors by bombarding stable sodium with neutrons.

$$^{23}_{11}\text{Na} + ^{1}_{0}\text{n} \rightarrow ^{24}_{11}\text{Na} + \gamma\text{-rays} \qquad 10.2$$

As the neutron is uncharged it can readily enter the ^{23}Na nucleus to produce a different atomic species, $^{24}_{11}$Na. This isotope is unstable and it reaches the stable state by emitting a highly energetic β-particle (1.39 MeV) and two photons of energies 1.37 and 2.75 MeV respectively.

$$^{24}_{11}\text{Na} \rightarrow ^{24}_{12}\text{Mg} + \beta^- + \gamma\text{-rays} \qquad 10.3$$

^{24}Na is unstable because of an excess of neutrons, and one of these in effect decays to give a proton and an electron, leaving the element Mg in place of Na.

Table 10.2 gives details of some of the radioisotopes in daily use in laboratories throughout the world, and their applications range from the study of ion transport across cell membranes (^{22}Na, ^{42}K) to the treatment of thyroid tumours (^{131}I).

Table 10.2. Physical data of radioisotopes. The half-life of a radioactive species is the time required for the radioactivity of a given amount of the element to decay to half its original value.

Isotope	Half-life	Radiation	Energy
^{45}Calcium	165 days	β^-	250 KeV
^{14}Carbon	5760 yrs	β^-	160 KeV
^{3}Hydrogen (Tritium)	12 yrs	β^-	18 KeV
^{131}Iodine	8 days	β^-	610 KeV
		γ	360 KeV
^{32}Phosphorus	14 days	β^-	1.7 MeV
^{40}Potassium	1.3×10^9 yrs	β^-	1.3 MeV
		γ	1.5 MeV
^{42}Potassium	12 hrs	β^-	2 and 3.6 MeV
^{22}Sodium	2.6 yrs	β^+	0.51 MeV
		γ	1.28 MeV
^{35}Sulphur	87 days	β^-	0.167 MeV

The unit of radioactivity for an isotope is basically the number of disintegrations per second and this was chosen as it is proportional to the number of radioactive atoms present. The activity unit, called the *Curie*, is that of a gram of radium and is equal to to 3.7×10^{10} disintegrations per second.

10.4 Interactions of Radiations with Matter

Because they are highly energetic, radiations interact very strongly with atoms and molecules in their path. They have in fact sufficient energy to cause *ionization* and so are often referred to as *ionizing radiations.* The end product of this ionization is radiation damage in the material which the atoms and molecules comprise. If the material is an animal cell, however, then the end result could be mutilation or death for the animal.

X-ray photons are absorbed by materials by transferring their energy to atomic electrons in much the same manner as light photons, except in the case of X-rays, the electrons gain sufficient energy to leave the atom together. The same absorption law holds for both cases

$$I_x = I_0 \exp(-\mu x) \qquad 10.4$$

where I_0 is the intensity of X-rays at the surface of the material and I_x is the intensity at some distance x from the surface. μ, the absorption coefficient of the material, increases with the *atomic number*, i.e. number of protons in the nucleus, of the material. The more dense the material, the greater is the attenuation of original intensity and as bones are much denser than the surrounding tissue, they are well defined on X-ray photographs. The depth of penetration of X-rays increases with energy (Table 10.1).

γ-photons have more than one mode of interaction with matter. The rays can ionize an atom in much the same way as X-rays, or they can penetrate the nucleus if they have sufficient energy. When this occurs the γ-ray energy is taken up in producing an electron and its antiparticle, the positron. These particles leave the nucleus and create further ionizations. In this case, the energy of the γ-photon creates two particles and this awesome feat is reversed when a positron and an electron collide – they are annihilated and γ-photons are produced.

The charged particles commonly encountered (γ- and β-particles and positrons) interact with the electric fields of the external atomic electrons and again the latter can be given sufficient energy to leave the atom and produce a so-called *ion pair* consisting of an electron and an ionized atom.

Neutrons are very damaging. Because they are uncharged, they can easily penetrate the nucleus, render it unstable, and so produce further radiations.

10.5 Biological Effects of Radiation

The principal material of which most cells are composed is of course water and much effort has been expended in order to elucidate the effect of radiations on water. It now appears that when the primary ion pairs are produced, reactive ions and free radicals are formed in the water

$$H_2O + \text{absorbed energy} \rightarrow H_2O^+ + e^- \qquad 10.5$$

$$H_2O + e^- \rightarrow H_2O^- \text{ (solvated electron)} \qquad 10.6$$

It is proposed that the highly reactive species so formed then interact with, and indeed, damage other molecules, e.g. DNA, which may not have been themselves hit by the ionizing radiations.

The main unit used to quantify the interaction of radiation with matter is the *RAD* (Radiation Absorbed Dose) and this is the quantity of radiation that will result in an energy absorption of 10^{-2} J per kg of tissue.

As far as radiation damage is concerned, not only is the total energy absorbed by the tissue important, but also how localized is the energy absorption. It appears that densely ionizing particles are the more damaging and this is probably because in many cases several hits are required on one site in a molecule to damage it permanently. The *linear energy transfer* (LET) of a particle determines the radiation damage, and the LET is the energy deposited per metre of particle path length. It has been found that LET increases as the square of the particle charge and decreases as the square of the velocity, so α-particles are relatively more damaging than β-particles, while low energy β's are more damaging than high energy β's.

The LET determines the *relative biological effectiveness* (RBE) of a radiation type. The RBE of γ and X-radiation is about equal and taken as 1. The RBE of fast moving β particles is also 1, while that of very slow-moving β's is 2. The RBE value for slow neutrons and α-particles are 3 and 10 respectively. This means that if 1 rad of γ-rays produces a certain effect in an animal or tissue, then only 0.5 rads of slow β's or 0.1 rads of α-rays will produce the same effect.

However, even with the RBE of the various radiations taken into account, we cannot simply state the biological effect of a dose of, say, 1 rad. This is because some animals are more radiosensitive than others, just as different tissues also vary. The total body dose (X- or γ-rays) that will kill 50% of a certain population of animals (the so-called LD_{50}) is 3000 rads for newts and 250 rads for guinea pigs, whereas the LD_{50} for man, computed from the Nagasaki and Hiroshima data is about 650 rads. If certain very radiosensitive parts of the body, notably the spleen, are shielded, then the LD_{50} increases.

However, even taking species and tissue variation into account, we still have not solved the problem of assessing the effect of a certain radiation dose, because not only is the total dose important, but also how it is administered. Table 10.3 shows the total dose necessary to produce in man a fixed intensity of skin reddening at two weeks following X-radiation. Narrow beams were used, so the total irradiated area was very small. The

Table 10.3. (From Alexander, 1965).

Dose rate (rad/min.)	Irradiation time	Total dose (rad)
500	1 min.	500
50	15.5 min.	780
5	260 min.	1300
$\frac{1}{2}$	4500 min.	2250

fact that the required total dose increased markedly as the dose rate decreased implies that there are recovery processes in the skin.

Skin reddening, erythema, is just one of the many gross effects produced by radiation. Relatively low doses, as little as 25 rads (X-ray equivalent), will produce a significant drop in the white cell count of the blood and so, if there is any doubt about possible exposure to radiation, a simple count should be carried out. The red blood cells on the other hand are more radioresistant.

Sterility in men occurs after an exposure to about 4000 rads while in women it occurs after 1000. If 400 rads are received, then men remain fertile for many weeks, are then sterile for several months, and finally recover fertility once more. The reason for this is that sperms are resistant, but the sperm precursor cells are not and the months of sterility are ended only by the repopulation of the testes with sperm precursor cells. For anyone who is worried by this, the doses likely to be encountered during an X-ray examination are tabulated (Table 10.4).

Table 10.4. Medical radiation doses. The dose rates are for modern, well-shielded machines.

	Dose to chest	Dose to gonads
Hospital Chest X-Ray	5 - 30 mR	0.1 - 1 mR
Mass Radiography	100 - 500 mR	1 - 5 mR

Great care has to be taken in the radiological examination of women of child-bearing age because during a radiological examination of the pelvic region itself, the dose to the gonads can exceed 1 rad. Experiments with mice, for example, have shown that if 200 rads are received to the pelvic region within a few days after conception then no live births occur as the zygote is very radiosensitive. However, if the embryo is irradiated during the differentiating stages, then the birth of live but abnormal offspring will result and as little as 5 rads can produce abnormalities. Irradiation in the latter stages of pregnancy produces very few abnormalities.

From the above observations it is possible to draw the following conclusions about the relative radiosensitivity of different cells in the body: the most radiosensitive cells are those which (i) have the highest division rate, (ii) retain the capacity for division the longest, (iii) are the least differentiated.

However, not only do radiations induce gross changes in tissues, but they also produce *mutations* in the genetic material of the animal and these changes may only be manifest generations later. For radiation death, there is clearly a lower limit below which it is unlikely for any deaths to occur (Fig. 10.1a). However, careful experiments performed on large numbers of mice seem to show that the number of radiation-induced mutations (Fig. 10.1b) increases linearly with dose and so genetically any exposure to radiation can be considered harmful.

As X-ray photographs must be taken and we must work with radioactive materials, we have to decide on certain maximum permissible levels. In considering genetic ef-

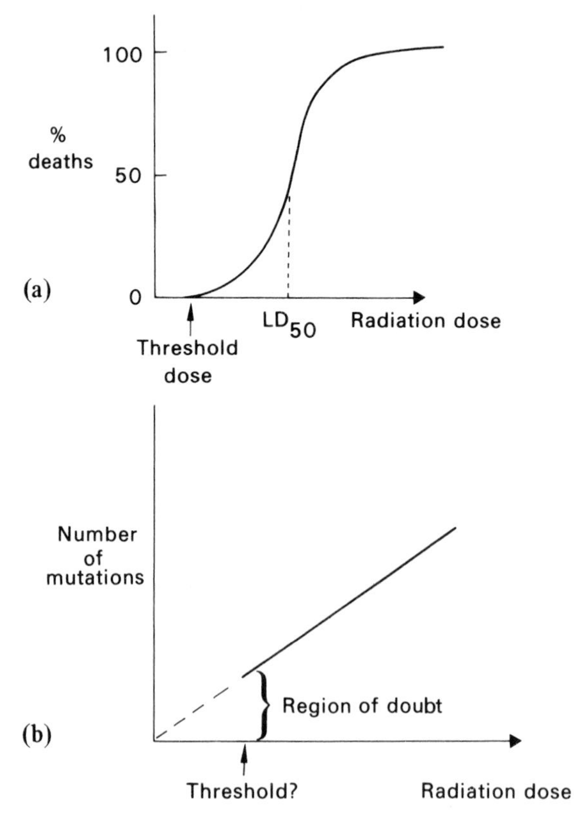

Fig. 10.1. (See text for explanation)

fects, we are not concerned with individuals, but rather the average dose to the whole population. The natural background level of 0.1 rad per year is the normal exposure to the population in Britain for example (Table 10.5), and in fact there appears to be not detectable increase in mutations in those regions, e.g. Kerala in India, where the

Table 10.5. Causes of natural background radiation and the average background dose (m rad/year) to the individual in Britain. In certain parts, e.g. Aberdeen, where most of the good, solid buildings are composed of granite, the background from soil and buildings is almost twice as high due to the presence of radioactive isotopes in granite.

Bomb Test fall out	5
Cosmic Radiation	30
Radiation from soil and buildings	20
Naturally ocurring isotopes, principally ^{40}K	20
Diagnostic X-rays	25
TOTAL	100 mRad/year

background rate is at least twice normal. Thus it seems that a population can tolerate a further 0.1 rad per year or, over a generation time of 30 years, 3 rads per generation.

A few special workers are permitted a greater exposure, but they contribute so little to the gene pool that the sum effect may be neglected. The main threat to our genetic heritage at present comes not from isotope users, atomic energy authorities, or even bomb tests, but from the therapeutic and diagnostic uses of X-rays. Fortunately this has been recognized for a long time and in the U.K. at least, the total dose to the population from medical X-rays adds only 20% to the natural background.

10.6 Radiation Detectors

Now that we know some of the possible dangers involved in handling isotopes and X-ray sources we are better equipped to plan experiments involving isotopes. The actual steps that should be carried out before and during the experiment to ensure that minimum risk is involved to oneself and one's colleagues are described in great and good detail by Thornburn (1972) and as they involve more common sense than science they will not be repeated here. However, an important part of the planning of an experiment will be described, and that is the background knowledge required to choose the most efficient way of assaying the isotopes (and X-ray sources) used in the experiment. All of the various techniques will record background radiation (Table 10.5) even when there is no radioactive sample in the machine and so this background has to be subtracted form the total counts with the sample in place.

(i) **Ionization Chamber and Geiger-Muller Tube**

One of the earliest forms of counting devices was the ionization chamber, which is still widely used in hospitals to monitor the radiation from X-ray machines. The chamber (Fig. 10.2) is normally filled with air and has two fixed parallel electrodes

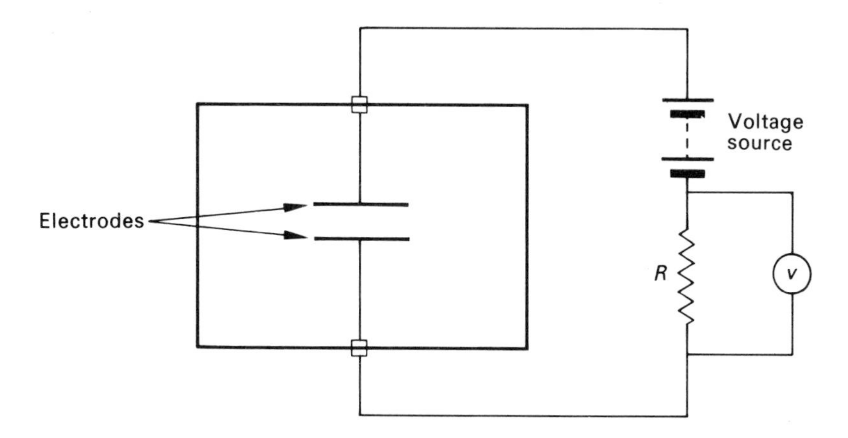

Fig. 10.2. The ionization chamber. A steady beam of radiation produces a constant current through the large resistance R. The resulting voltage is read on a high resistance voltmeter V.

across which can be applied a potential difference of several hundred volts. Normally the air inside the chamber does not conduct electricity and so no current will flow through the system. However, when radiations pass through the chamber they will ionize the enclosed gas; the electrons will move to the anode, and the positively charged ions to the cathode. If a steady beam of radiation passes through the chamber, then a steady current will flow through the resistance R and its magnitude can be calculated from the reading on the high impedance voltmeter V. When X-rays are the radiation source, the dose rate D in Röntgens per hour (problem 10.1, see below) is directly related to the current i by the equation

$$i = 9.25 \times 10^{-14} \times VD \text{ amps} \qquad 10.7$$

where V is the volume of the chamber in cm^3. If the voltmeter has a suitably short time-constant the chamber can be used as a proportional counter to count single particles as each of these will produce a single voltage pulse.

The sensitivity of such an ionization chamber can be increased by raising the potential difference between the plates (HT in Fig. 10.3) to about 1 kV and by filling the chamber with special gases. The modifications were introduced by Geiger and Muller in 1920 and in modern versions of the G-M tube the gases are normally argon and ethyl alcohol and a thin central wire acts as an anode while the metal of the chamber (earthed) is the cathode. The radiations enter through a thin metal or mica window at one end so that absorption before entering is reduced (Fig. 10.3). When an ion pair is pro-

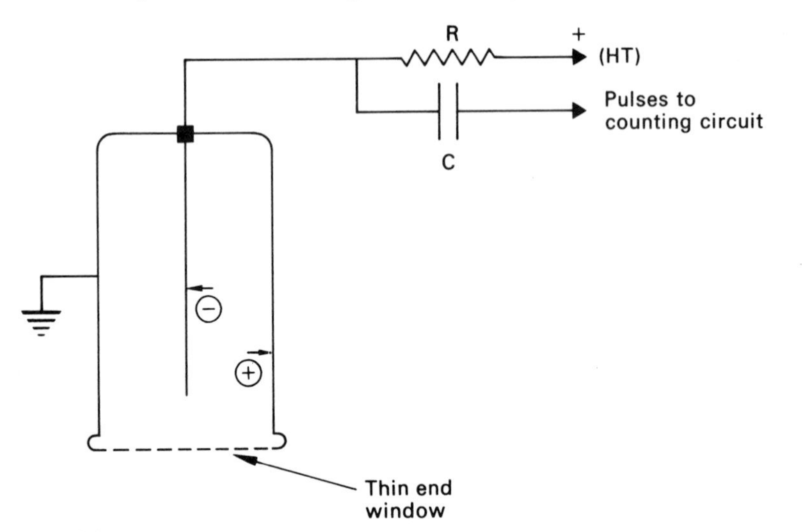

Fig. 10.3. The Geiger-Muller tube. As the G-M tube is normally used to count single pulses the time constant of the circuit RC must be short.

duced the electron moves with a high velocity towards the central wire and the positive argon ion moves slowly to the cathode. The fast moving electron ionizes further argon atoms and an electronic avalanche is set up, the size of which is independent of the energy and position of the initial ionization. This results in an anode current. As the electrons from the avalanche are collected, the slowly moving argon ions act as an electrostatic screen and this reduces the field at the wire to a value below that necessary for ionization by collision, so that the discharge should cease. However, there is the possibility that the positive argon ions can eject electrons from the cathode and to prevent this the alcohol is added to quench the discharge, and it does so because of its low ionization potential. The argon ions on their journey to the cathode are neutralized by acquiring an electron from the alcohol molecules. The ions reaching the cathode are then alcohol ions and although they themselves are neutralized at the cathode, the energy is absorbed in dissociating the alcohol molecule rather than in producing further electrons from the cathode. The discharge thus ceases when the field around the anode has fallen sufficiently.

The advantages of the G-M tube lie in its sensitivity and in its operating characteristics (Fig. 10.4). The count rate is constant over a relatively large range of applied

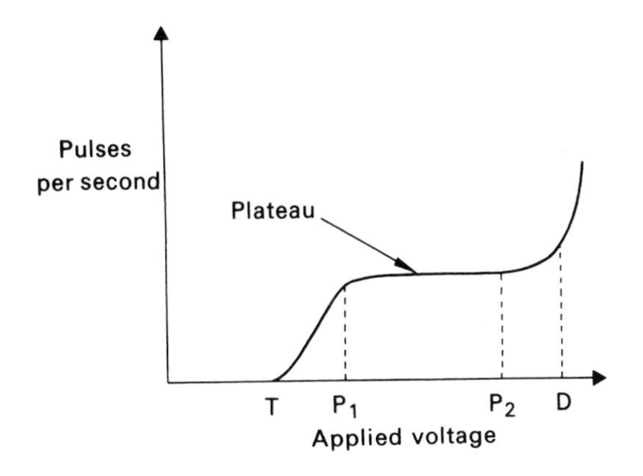

Fig. 10.4. Operating characteristics of the G-M tube. The tube starts to count at T, the threshold voltage. The operating region (or G-M plateau) is between the voltage at P_1 and P_2. At higher voltages, D, continuous discharge occurs (and the tube may be ruined).

voltages. Although the G-M tube can be used to count relatively low energy particles, e.g. β from ${}^{14}_{6}C$, it is useless for the β's from tritium (${}^{3}_{1}H$) as these have insufficient energy to penetrate even the thin mica window. The G-M tube is also inefficient at counting γ-rays and for these a different counting technique is employed.

A serious disadvantage of the G-M tube is that it has a relatively long insensitive time following each current pulse. This is made up of a dead time during which the electric field near the anode has dropped below the counting threshold and a recovery time during which pulses of reduced size are produced. The insensitive time, approximately 300 μs, is usually increased and set at a definite value by the addition of an external electrical circuit. At high counting rates a correction for lost counts must be applied.

Problem 10.1

(*a*) Given that 1 Röntgen is defined as that dose of radiation which produces 2.1×10^9 ion pairs in 1 cm^3 of air, derive the relationship between current i and dose rate D given in equation 10.7.

$$1 \text{ electronic charge} = 1.6 \times 10^{-19}\ \text{C}$$

(*b*) The rad is the radiation absorbed dose and is defined as an energy absorption of 10^{-2} J per kg material. What is the dose absorbed by air exposed to 1 Röntgen of X-rays?
(34 eV is associated with 1 ion pair; density of air = 1.29×10^{-3} gm cm^{-3})

(ii) **Scintillation Counters**

One of the earliest observations of nuclear physics was that certain materials emit absorbed radiation in the form of tiny light pulses or scintillations. The first scintillation counter consisted of a zinc sulphide screen viewed by a low power microscope. It was found that for accurate measurements of activity with this system, the scintillation counts had to be corrected for a dead time, which in this case was the blink time of the observer. The design of a modern scintillation counter is shown in Fig. 10.5.

In one form of scintillation counter a crystal, usually composed of sodium iodide and activated by thallium, is used. The radiation produces light pulses along its track and the pulses are counted using a photomultiplier tube. The impinging light photons drive electrons from the surface of a photocathode at the first stage of a photomultiplier. The initial pulse of electrons is amplified in the photomultiplier which consists of a series of electrodes. Application of an increasing positive potential between successive electrodes accelerates the electrons from one electrode to the next and at each impact 2–5 secondary electrons per primary may be produced. The final highly amplified current is collected at the anode and on passing through a suitable resistor can produce a pulse of several volts.

These counters have certain advantages over the G-M tube, as they have a very short insensitive time and also they count γ-rays much more efficiently, because of the greater density of the sensitive material. The photomultiplier does however require a higher and much more stable operating voltage. In another form the radioactive sample is mixed with a liquid scintillator material, which has two main components: the solvent, usually xylene and the primary solute, often diphenyloxazole. Most of the energy is absorbed from the radiation by the solvent and then transferred to the solute, which actually emits the light. Often a small amount of secondary solute is added to shift the wavelength of the emitted photons in order to increase the transparency of the scintillator to the light photons. As the radioactive compounds are dissolved in the scintillator the counting efficiency, especially for particles of low energy (β-particles from ^{3}H and ^{14}C) is relatively high. The sample is usually placed over a photomultiplier.

As the magnitude of the light flashes, and hence of the final output, depends on the energy of a particle, the final electronics can be arranged so that only pulses of a certain height are counted. With several electronic discriminators or gates set at different levels, a corresponding number of different isotopes can be counted in the same bottle, provided their radiations have sufficiently different energies, e.g. ^{3}H, ^{14}C, ^{32}P.

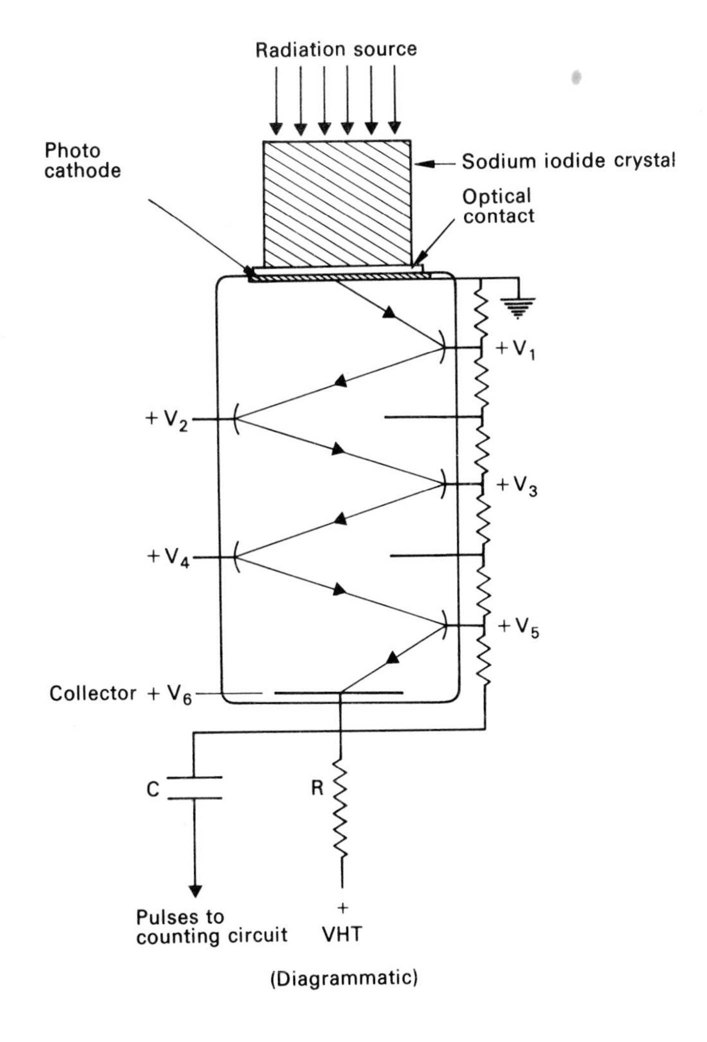

Fig. 10.5. Crystal scintillation counter. The crystal and photomultiplier are built into a light-proof box and the photomultiplier is surrounded by a Mu metal screen to prevent interference from magnetic fields.

(iii) Cerenkov Counting

Machines that assay by liquid scintillation method can also be used to count radioactive samples in aqueous solution without any scintillator, provided that the particles emitted are charged and are moving with sufficiently high velocities. When the velocity of the particle is greater than the velocity of light in the medium in which the particle is travelling, light photons are produced as a result of an electromagnetic shock wave induced in the medium in which the particle is travelling (analogous to a supersonic bang). The isotopes ^{22}Na, ^{24}Na and ^{42}K, widely used in biology, can be assayed by this technique as they emit β-particles of a sufficiently high energy.

(iv) Autoradiography

To satisfy the needs of nuclear physicists, photographic emulsions have been developed that have a small grain size and low background on developing. Such films are now being widely used in biology to locate small numbers of specific atoms or molecules in biological tissues. This technique gives the highest resolution with particles of low energy (β-particles from ^{3}H and ^{14}C) where the length of track of silver atoms comprising the latent image is short. On developing, the latent images are amplified and appear as black grains of metallic silver.

In this technique, a thin slice of tissue that has previously been treated with an appropriate radioactive substance is placed over the film (Fig. 10.6) and left for a

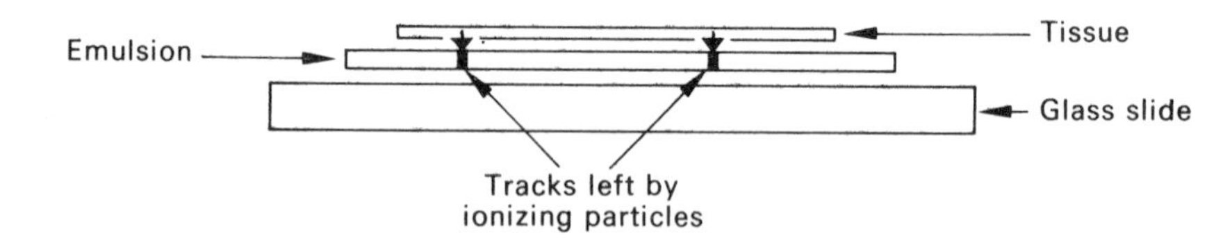

Fig. 10.6. Autoradiography. When the emulsion is developed, the tracks are black against a lighter background.

suitable, often very prolonged, time. After development, the film can be examined under the optical or electron microscope to determine the distribution of radioactive substance in the tissue.

Young (1969) has used autoradiography to show that there is a continual turnover of discs in the outer segments (rods) of vertebrate photoreceptor cells (Fig. 10.7). After a short exposure (pulse labelling) to a mixture of amino acids injected into the blood, it can be seen by sacrificing animals at different times that the amino acids are incorporated into the disc membranes (separate experiment show that it is incorporated into the protein part of the photopigment rhodopsin) starting at the discs near the inner segments. These labelled discs are gradually pushed upwards by non-labelled discs until they reach the pigment epithelium where they are broken

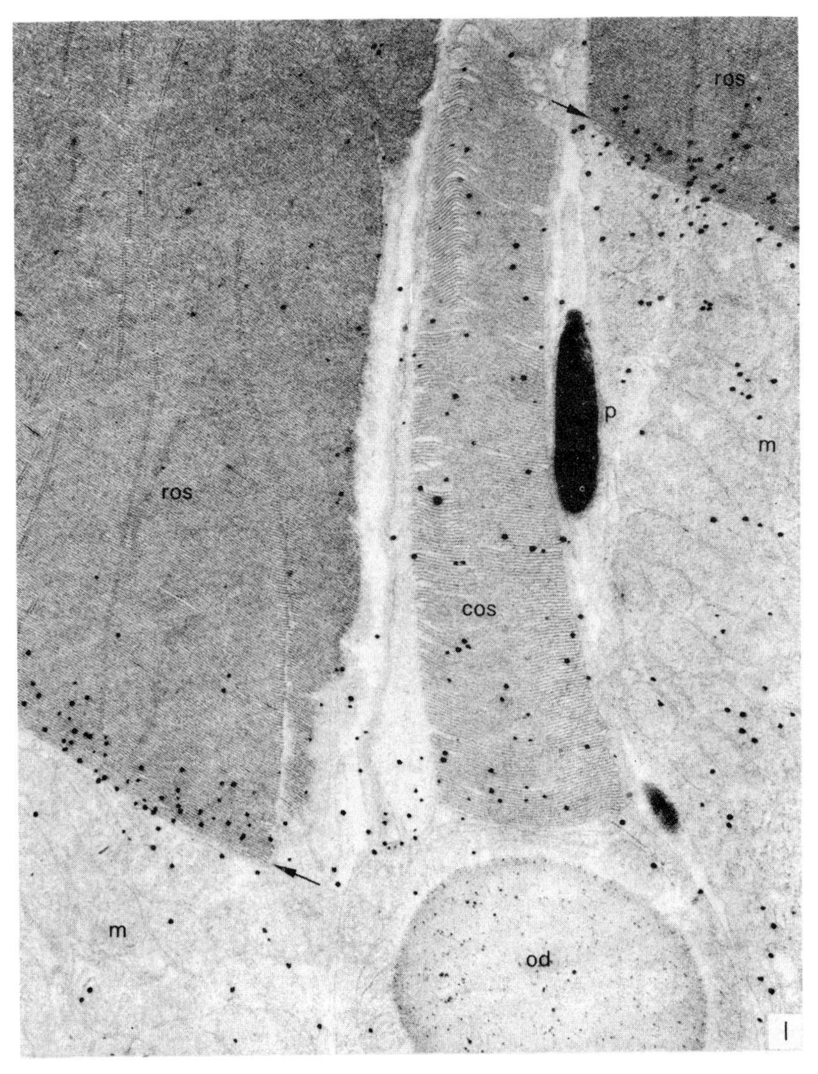

Fig. 10.7. One day after injection. The diffuse distribution of renewal protein in the principal cone outer segment (cos) contrasts markedly with the discrete and heavy labelling of a small group of discs near the base of the rod outer segment (ros). The tracks appear black as only silver grains remain on the emulsion after the developing process. Pigment granules (p) surrounding the cells, mitochondria (m) in the inner segments of the rods, and coloured oil droplets (od) at the base of the cones can also be clearly seen. (Electron micrograph, autoradiogram x 14,000). (This electron micrograph was kindly supplied by Dr. R.W. Young of the University of California at Los Angeles and is used with his permission.)

down. Cones, on the other hand, are diffusely labelled from the beginning, suggesting that all regions of the membrane are equally accessible to the source of label in the inner segment.

References

Alexander P. (1965) *Atomic Radiation and Life.* Pelican.
Coggle J.E. (1971) *Biological Effects of Radiation.* Wykeham, London.
Epstein H.T. (1963) *Elementary Biophysics.* Addison-Wesley, Reading, Mass.
Glasstone S. (1958) *Sourcebook on Atomic Energy.* vanNostrand, Princeton.
Report of the International Commission on Radiobiological Units and Measurements (1959). National Bureau of Standards, Washington D.C.
Setlow R.B. & Pollard E.C. (1962) *Molecular Biophysics.* Pergamon, London.
Thornburn C.C. (1972) *Isotopes and Radiation in Biology.* Butterworths, London.
Young R.W. (1969) A Difference between Rods and Cones in the Renewal of Outer Segment Protein. *Investigative Ophthalmology* 8,222-231.

Appendices

Appendix I. The International System of Units (S.I.)

Table A.1. Defined S.I. Units.

Physical quantity	Name of Unit	Symbol for Unit
Mass	kilogram	kg
Length	metre	m
Time	second	s
Electric current	ampere	A
Thermodynamic Temperature	kelvin	K

Table A.2. Special names and symbols for SI derived units

Physical quantity	Names of SI Unit	Symbol for SI Unit	Definition of SI Unit	Equivalent form
energy	joule	J	$kg\ m^2\ s^{-2}$	Nm
force	newton	N	$kg\ m\ s^{-2}$	$J\ m^{-1}$
pressure	pascal	Pa	$kg\ m^{-1}\ s^{-2}$	$N\ m^{-2}$
power	watt	W	$kg\ m^2\ s^{-3}$	$J\ s^{-1}$
electric charge	coulomb	C	As	A s
electric potential difference	volt	V	$kg\ m^2\ s^{-3}\ A^{-1}$	$J\ A^{-1}\ s^{-1}$, $J\ C^{-1}$
electric resistance	ohm	Ω	$kg\ m^2\ s^{-3}\ A^{-2}$	$V\ A^{-1}$
electric capacitance	farad	F	$A^2\ s^4\ kg^{-1}\ m^{-2}$	$A\ s\ V^{-1}$, CV^{-1}
magnetic flux	weber	Wb	$kg\ m^2\ s^{-2}\ A^{-1}$	V s

Table A.3. Units to be allowed in conjunction with SI.

Physical quantity	Name of unit	Symbol for unit	Definition of unit
volume	litre	l	$10^{-3}\ m^3 = dm^3$
radioactivity	Curie	Ci	$37 \times 10^9\ s^{-1}$
energy	electron volt	eV	1.6×10^{-19} J

Table A.4. Examples of units contrary to SI with their equivalent.

Physical quantity	Name of unit	Equivalent
Length	ångström	10^{-10} m
Force	dyne	10^{-5} N
Pressure	atmosphere	10^{5} Nm^{-2}
	torr	133 Nm^{-2}
Energy	erg	10^{-7} J
	calorie	4.2 J

Table A.5. Prefixes for SI Units. These prefixes may be used to construct decimal fractions or multiples of units.

Multip

Fraction Multiple	Prefix	Symbol	Multiple	Prefix	Symbol
10^{-1}	deci	d	10	deca	da
10^{-2}	centi	c	10^{2}	hecto	h
10^{-3}	milli	m	10^{3}	kilo	k
10^{-6}	micro	μ	10^{6}	mega	M
10^{-9}	nano	n	10^{9}	giga	G
10^{-12}	pico	p	10^{12}	tera	T

Table A.6. SI Supplementary Units.

Physical quantity	Name of SI Unit	Symbol for SI Unit
Plane angle	Radian	rad
Solid angle	Steradian	sr

References

Quantities Units and Symbols (1971). Published by the Royal Society, 6 Carlton House Terrace, London.

van Assendelft O.W., Mook G.A. & Zijlstra W.G. (1973) International System of Units (SI) in Physiology. *Pflügers Arch,* **339,** 265-272.

Appendix II Vector Algebra

Vector addition has been dealt with in Chapter 1; here we shall deal with *vector multiplication.*

(a) The *scalar product* of two vectors **A** and **B** is defined by the relationship

$$\mathbf{A} \cdot \mathbf{B} = A\,B \cos\theta$$

where θ is the smaller of the two possible angles between the vectors (Fig. A.1).

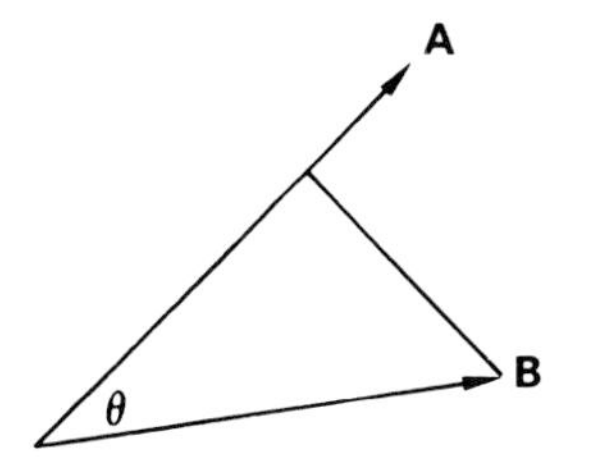

Fig. A.1. The projection of **B** on **A** has magnitude $B \cos\theta$ and this multiplied by the magnitude of **A** gives the scalar product $AB \cos\theta$.

With this definition of scalar product, a number of important physical quantities can be described as the scalar product of two vectors, e.g. mechanical work (Chapter 1) and electrical potential (Chapter 9).

(b) The *vector product* of two vectors **A** and **B** is written as **A** x **B** and is another vector **C** where the magnitude of **C** is given by

$$C = A\,B \sin\theta$$

where θ is the angle between **A** and **B** (Fig. A.2). The direction of **C** is perpendicular to the plane of the two vectors **A** and **B** and is defined as follows:

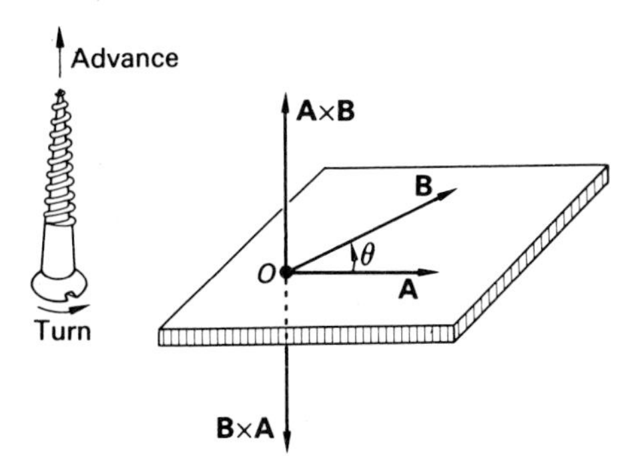

Fig. A.2. Vector products of **A** and **B**. The product **A** x **B** points upward, while the product of **B** x **A** points downward. The magnitude of both products is $AB \sin\theta$, but **B** x **A** = − **A** x **B**.

Imagine a right-hand screw with its axes perpendicular to the plane of **A** and **B** and let the screw be turned in the same way the first vector **A** turns when it is rotated through the smallest angle θ that will bring it parallel with the second vector **B**. The product **A** x **B** then points in the direction of advance of the screw. The vector product **B** x **A** has the same magnitude as **A** x **B** but points in the opposite direction.

An example of a quantity that is a vector product is the force on a moving charge in a magnetic field (Chapter 9).

Answers to Problems

Problem 1.1 (p8)

Answer in text.

Problem 1.2 (p10)

(a) Kinetic Energy $= \frac{1}{2}\ mv^2$
$= 8.8 \times 10^2$ J

Power $= 8.8 \times 10^2\ Js^{-1}$
$= 0.88$ kW

(b) deceleration $= v/t$
$= 5\ ms^{-2}$
≈ 0.5 g

Problem 1.3 (p20)

(a) Apply equation 1.8b, with $v_t = 0$

i.e. $v_o^2 = 2as$

$\therefore a = 14.3\ ms^{-2}$

Now, deceleration due to gravity $= 9.8\ ms^{-2}$

$\therefore$ deceleration due to air resistance $= 4.5\ ms^{-2}$

Which is approximately half the gravitational deceleration

(b) acceleration $= v/t$
$= 1000\ ms^{-2}$

which is approximately 100 g

(c) Kinetic energy $= 0.23 \times 10^{-6}$ J

(d) Flight is powered over a period of 10^{-3} s, hence power input required is 0.23×10^{-3} W. Maximum output from muscles $= 5.4 \times 10^{-6}$ W, which is clearly insufficient to power the jump.

(e) Answer in text.

Problem 2.1 (p30)

(a) This is known as the 'greenhouse effect' and it arises because the illuminated contents of the pack will reradiate longer wavelength radiation than that absorbed (equation 2.9). Because plastics (and glass) absorb these long wavelength radiations, the contents will ultimately heat up.

(b) A metal foil pack with a description or picture of contents would be all that is required, but this would be unpopular with sceptical shoppers. An alternative pack could have a transparent window on the bottom face, with an instruction on the pack to place that face downwards.

Problem 2.2 (p30)

Answer in text.

Problem 3.1 (p34)

(a) 1.3×10^4 Nm^{-2}
(b) i. 0.8×10^4 Nm^{-2}
(b) ii. 1.8×10^4 Nm^{-2}

Problem 3.2 (p42)
Answer in text.

Problem 4.1 (p47)

Apply equation 4.1 to the data to give the answer 'yes'.

Problem 4.2 (p48)

(a) Linear velocity of flow = volume flow/area
= 28×10^{-2} ms^{-1}
Kinetic energy density is given by $\frac{1}{2}\rho v^2$
= 45 Nm^{-2},
which is small compared to the arterial pressure of 1.3×10^4 Nm^{-2}.
(b) The pressure associated with a linear velocity of flow of 84×10^{-2} ms^{-1} is 360 Nm^{-2} which is certainly sufficient to explain the observed pressure difference.
(c) Kinetic energy created = 10^3 Nm^{-2}.

Problem 4.3 (p49)

From the continuity equation, the velocity of flow increases by a factor of 5 when the lumen narrows to under one fifth. Therefore the kinetic energy density of the fluid goes up from 40 Nm^{-2} to 10^3 Nm^{-2} and the arterial pressure is reduced by this amount.

Problem 4.4 (p55)

From equation 4.16, $N_R = 1500$, which is less than the critical number.
During heavy exercise, when the velocity increases five-fold, then $N_R = 7500$, which is greater than the critical number.

Problem 4.5 (p55)

(a) Linear velocity of flow in lumen = 2.3×10^{-4} ms^{-1}
Linear velocity of flow in pores = 4.6×10^{-4} ms^{-1}
Pressure drop across a 195μ length of lumen = 7.1×10^2 Nm^{-2}
Pressure drop across sieve plate = 9.2×10^2 Nm^{-2}
Total pressure drop across 200 μ length = 16.3×10^2 Nm^{-2}
Total pressure drop across 1 metre = 7.85×10^6 Nm^{-2}
(b) The total possible osmotic driving force would be given by

$$\Delta\pi = RT\Delta C_s$$

providing the phloem membranes at A were impermeable to sucrose

i.e. $\Delta\pi = 7.5 \times 10^5$ Nm^{-2} (assuming $T = 293°$ K)

The osmotic driving force is clearly insufficient to drive the sucrose solution through the phloem at the given rates.
(c) Almost every possible driving force has been suggested at some time and at present the only non-controversial statement in this troubled field is that there is as yet no consensus of opinion.

Problem 5.1 (p63)

From equation 5.6, $h = 0.7$ m (maximum)
Other possible mechanisms are root pressure (osmotic driving force due to a gradient of solute concentration across root cell membranes), and transpiration.

Problem 5.2 (p68)

Upward surface tension force $= 42 \times 10^{-5}$ N
downward gravitational force $= 24.5 \times 10^{-5}$ N
When the surface tension is reduced by detergents,
upward force $= 24 \times 10^{-5}$ N.

Problem 6.1 (p82)

(a) Answer given in text
(b) 10%

Problem 6.2 (p82)

(a) The energy (E') falling on an object a distance r from the source is inversely proportional to r^2

$$\text{i.e. } E' = \frac{kE}{r^2}$$

Suppose a fraction of this energy is reflected from the object

$$\text{i.e. } E'' = \frac{kfE}{r^2}$$

Again from the inverse square law, the proportion of this energy falling on the source will be inversely proportional to r^2

$$\text{i.e. } E''' = \frac{kE''}{r^2}$$

(assuming the bat moves a negligible distance during the process). Hence the intensity of the echo is inversely proportional to r^4.
(b) Ratio of echo intensities from small object $= (r_1/r_2)^4$
where $r_1 = 0.3$ m and $r_2 = 0.25$ m
Hence $I_2/I_1 = 2.06$
Similarly, ratio of echos from large object = 1.2

Problem 6.3 (p82)

(a) Suppose that v is the relative velocity of the bat and object. Distance between bat and object at first echo $= (\tau + t)v$
and distance at second echo $= tv$
Now, ratio of intensity of second echo to that of first is $1 + \alpha$ and from problem 6.2 this is given by

$$1 + \alpha = \left(\frac{(\tau + t)v}{tv}\right)^4$$

$$\text{or } t = \frac{\tau}{(1 + \alpha)^{\frac{1}{4}} - 1}$$

(b) $t = 5s$

Problem 6.4 (p83)

The sound heard directly is 42×10^3 Hz as in this case there is no relative motion of the source and observer. The echo, however, will have a different frequency. The wall receives and reflects a note f' Hz where

$$f' = \frac{3.3 \times 10^2}{3.3 \times 10^2 - 10} \times 42 \times 10^3$$

Now the bat approaches this source and therefore perceives a note of frequency f'' where

$$f'' = f' \times \frac{3.3 \times 10^2 + 10}{3.3 \times 10^2}$$

It therefore hears two frequencies 42×10^3 Hz and 45×10^3 Hz.
The bat will also hear a series of beats when the two frequencies are combined and the beat frequency will be the difference between the two.

Problem 7.1 (p93)

(a) Applying the lensless eye equation (7,3) with $\mu_1 = 1$,
$\mu_2 = 4/3$, $u = -\infty$ and $r = \frac{2}{3}R$
where R is the radius of the eye, then
$v = 2\frac{2}{3}R$ and the image is therefore behind the retina.
(b) Applying the thin lens equation (7.5) with
$u = +2\frac{2}{3}R$, $v = +R$, then
$f = +8R$ and so converging lenses are required.
(c) The lens acts as a filter for blue light (an effect which increases with age) and so he might tend to use blue more than before.

Problem 7.2 (p105)

The path difference between successive rays reflected from the wing cover = 0.8 cos (45°) μm = 565 nm. Rays of light of this wavelength will therefore constructively interfere at the eye and wing cover will therefore have a greenish hue.

Problem 8.1 (p137)

Change in optical density at 500 nm = 0.4
applying equation 8.8 with
$\epsilon_{mol} = 41{,}000$; $D = 0.4$; $x = 1$ cm then
concentration of rhodopsin = 10^{-5} ML^{-1}.
but, molecular weight = 40,000
$\therefore$ dry weight of rhodopsin present = 0.4 mg/ml.

Problem 9.1 (p153)

Answer in text

Problem 9.2 (p153)

Answer in text

Problem 9.3 (p156)

Applying equation 9.16

$V = \Sigma RI - \Sigma E$
$\therefore V = I(2R_e + R_m) - E$
but from equation 9.14
$I = E/(R_v + 2\,R_e + R_m)$
$\therefore V = -ER_v/(R_v + 2R_e + R_m)$
$\therefore V$ will only give a true reading of E when
$R_v \gg 2R_e + R_m$

Problem 9.4 (p164)

(a) Applying equation 9.24 with
$E_m = -60$ mV; E_{Na}; +55 mV and $E_K = -75.5$ mV
then $R_{Na}/R_K = 7.4$

Now $\dfrac{1}{R_K} + \dfrac{1}{R_{Na}} = \dfrac{1}{R_m}$

where R_m is the membrane resistance ($10^3\ \Omega\ \text{cm}^2$)
$\therefore R_K = 1.14 \times 10^3\ \Omega\ \text{cm}^2$
(b) When $E_m = +40$ mV, application of equation 9.24

gives $\dfrac{R_K}{R_{Na}} = 7.7$

If R_K is assumed constant during the action potential (i.e. at a value of $1.14 \times 10^3\ \Omega\ \text{cm}^2$), then $R_{Na} = 1.48 \times 10^2\ \Omega\ \text{cm}^2$. The equivalent membrane resistance in this case is 130 $\Omega\ \text{cm}^2$. The computed resistance is the larger because it does not take into account the fact that at the height of the action potential, the resistance of the potassium channel has already started to decrease.
(c) 1 μF = 10^{-6} CV^{-1}
Voltage change during action potential = 0.1 V
$\therefore$ For discharge, 10^{-7} C move through per cm^2 of membrane
$\therefore$ approximately 10^{-12} moles cm^{-2} move through.

Problem 9.5 (p169)

Applying equation 9.41
with $V = 600 \times 10^{-6}$ V, $L = 1 \times 10^{-3}$ m and $B = 2$ Wb m^{-2},
then $v = 300 \times 10^{-3}$ ms^{-1}
This velocity of flow is for an area of 0.07×10^{-4} m^2
$\therefore$ Volume flow is 2.1×10^{-6} $\text{m}^3\ \text{s}^{-1}$

Problem 10.1 (p180)

(a) 1 Röntgen/hour produces 0.93×10^{-13} amps cm^{-3}
If the volume of the chamber is V cm^3 and the dose rate is D
then $i = 0.93 \times 10^{-13}\ VD$ amps.
(b) 1 Röntgen = 0.88 rad.

Index